Solar Chimney Power Plant Generating Technology

Solar Chimney Power Plant Generating Technology

Edited by

Tingzhen Ming

*School of Civil Engineering and Architecture,
Wuhan University of Technology, Wuhan, P.R. China*
and
*School of Energy and Power Engineering,
Huazhong University of Science and Technology,
Wuhan, P.R. China*

AMSTERDAM • BOSTON • HEIDELBERG • LONDON
NEW YORK • OXFORD • PARIS • SAN DIEGO
SAN FRANCISCO • SINGAPORE • SYDNEY • TOKYO
Academic Press is an imprint of Elsevier

Academic Press is an imprint of Elsevier
125 London Wall, London EC2Y 5AS, UK
525 B Street, Suite 1800, San Diego, CA 92101-4495, USA
50 Hampshire Street, 5th Floor, Cambridge, MA 02139, USA
The Boulevard, Langford Lane, Kidlington, Oxford OX5 1GB, UK

Notices
Knowledge and best practice in this field are constantly changing. As new research and experience broaden
our understanding, changes in research methods, professional practices, or medical treatment may become
necessary.

Practitioners and researchers must always rely on their own experience and knowledge in evaluating and
using any information, methods, compounds, or experiments described herein. In using such information
or methods they should be mindful of their own safety and the safety of others, including parties for whom
they have a professional responsibility.

To the fullest extent of the law, neither the Publisher nor the authors, contributors, or editors, assume
any liability for any injury and/or damage to persons or property as a matter of products liability,
negligence or otherwise, or from any use or operation of any methods, products, instructions, or ideas
contained in the material herein.

ISBN: 978-0-12-805370-6

British Library Cataloging-in-Publication Data
A catalog record for this book is available from the British Library.

Library of Congress Cataloging-in-Publication Data
A catalog record for this book is available from the Library of Congress.

For Information on all Academic Press publications
visit our website at http://elsevier.com/

www.elsevier.com • www.bookaid.org

Publisher: Jonathan Simpson
Acquisition Editor: Simon Tian
Editorial Project Manager: Naomi Robertson
Production Project Manager: Nicky Carter
Designer: Matthew Limbert

Typeset by MPS Limited, Chennai, India

This book is dedicated to the memory of my beloved mother who died on March 30, 2012 before her 70th birthday.

Tenzhen Ming

Contents

CHAPTER 5 **Design and Simulation Method for SUPPS**
Turbines 127

Tingzhen Ming, Wei Liu, Guoliang Xu, Yanbin Xiong,
Xuhu Guan and Yuan Pan

CHAPTER 6 **Energy Storage of Solar Chimney 147**

Tingzhen Ming, Wei Liu, Chao Liu, Zhou Zhou
and Xiangfei Yu

Contributors

Dong Chen
School of Energy and Power Engineering, Huazhong University of Science and Technology, Wuhan, P.R. China

Renaud Kiesgen de Richter
Institut Charles Gerhardt Montpellier — UMR5253 CNRS-UM2 — ENSCM-UM1—Ecole Nationale Supérieure de Chimie de Montpellier, Montpellier, France

Tingrui Gong
School of Energy and Power Engineering, Huazhong University of Science and Technology, Wuhan, P.R. China

Xuhu Guan
School of Energy and Power Engineering, Huazhong University of Science and Technology, Wuhan, P.R. China

Jinle Gui
School of Energy and Power Engineering, Huazhong University of Science and Technology, Wuhan, P.R. China

Zhengtong Li
School of Energy and Power Engineering, Huazhong University of Science and Technology, Wuhan, P.R. China

Chao Liu
School of Energy and Power Engineering, Huazhong University of Science and Technology, Wuhan, P.R. China

Wei Liu
School of Energy and Power Engineering, Huazhong University of Science and Technology, Wuhan, P.R. China

Fanlong Meng
School of Energy and Power Engineering, Huazhong University of Science and Technology, Wuhan, P.R. China

Tingzhen Ming
School of Civil Engineering and Architecture, Wuhan University of Technology, Wuhan, P.R. China; School of Energy and Power Engineering, Huazhong University of Science and Technology, Wuhan, P.R. China

Tao Pan
School of Energy and Power Engineering, Huazhong University of Science and Technology, Wuhan, P.R. China

Yuan Pan
School of Electrical and Electric Engineering, Huazhong University of Science and Technology, Wuhan, P.R. China

Keyuan Peng
School of Energy and Power Engineering, Huazhong University of Science and Technology, Wuhan, P.R. China

Xinjiang Wang
School of Energy and Power Engineering, Huazhong University of Science and Technology, Wuhan, P.R. China

Tianhua Wu
School of Energy and Power Engineering, Huazhong University of Science and Technology, Wuhan, P.R. China

Yongjia Wu
School of Energy and Power Engineering, Huazhong University of Science and Technology, Wuhan, P.R. China

Yanbin Xiong
School of Energy and Power Engineering, Huazhong University of Science and Technology, Wuhan, P.R. China

Guoliang Xu
School of Energy and Power Engineering, Huazhong University of Science and Technology, Wuhan, P.R. China

Wei Yang
School of Energy and Power Engineering, Huazhong University of Science and Technology, Wuhan, P.R. China

Xiangfei Yu
School of Energy and Power Engineering, Huazhong University of Science and Technology, Wuhan, P.R. China

Yong Zheng
School of Energy and Power Engineering, Huazhong University of Science and Technology, Wuhan, P.R. China

Cheng Zhou
School of Energy and Power Engineering, Huazhong University of Science and Technology, Wuhan, P.R. China

Zhou Zhou
School of Energy and Power Engineering, Huazhong University of Science and Technology, Wuhan, P.R. China

Preface

The field of renewable and sustainable energy is changing rapidly worldwide, and various technologies concerning energy saving and renewable energy utilization are constantly being reported. Further, the widespread use of solar energy, as an alternate and nondepletable resource for agriculture and industry as well as other applications, is dependent on the development of solar systems which possess the reliability, performance, and economic characteristics that compare favorably with the conventional systems.

The solar chimney power plant system (SCPPS) or solar updraft power plant system (SUPPS), which is composed of the solar collector, the chimney, and the turbine, has been investigated all over the world since the German researcher Professor Jörg Schlaich made the brainchild in the 1970s. The SCPPS, due to its attractive advantages of being easier to design, more convenient to draw materials, higher operational reliability, fewer running components, more convenient maintenance and overhaul, lower maintenance expense, no environmental contamination, continuous stable running, and longer operational life span, has excited many researchers all over the world, especially in countries with plenty of deserts and arid and "useless" areas. However, the book on SCPPS written by Professor Jörg Schlaich was published 20 years ago, and now it is necessary to update the progress made in the state-of-the-art technologies of SCPPS over recent years worldwide.

In this book we are going to reveal the basic mechanisms of fluid flow, heat transfer, power output, energy storage, and operation procedure of the turbine of SCPPS. We hope this book can provide good guidance for developers who are interested in SCPPS.

The remaining chapters are arranged in the following way. In chapter "Introduction," we will present a brief introduction of the background of various solar energy power plant systems and SCPPS, and we will also introduce the recent research developments of SCPPS during the past 20 years. In chapter "Thermodynamic Fundamentals," basic theory related to thermodynamics and the efficiency of the SCPPS will be introduced. This covers the basic thermodynamic process, Brayton cycle, and exergy analysis of various SCPPSs. Chapter "Helio-Aero-Gravity (HAG) Effect of SC" unveils the Helio-Aero-Gravity (HAG) effect of the SCPPS. In this chapter, how the SCPPS operates will be analyzed in detail dealing with the various parameters including the dimensions and ambient. In chapter "Fluid Flow and Heat Transfer of Solar Chimney Power Plant," a mathematical model describing the fluid flow, heat transfer, and power output of the SCPPS will be presented, validation of the model by comparing the experimental results of the Spanish prototype will also be presented. Later, optimization of the dimensions of SCPPS based upon the results of power output will be performed.

In chapter "Design and Simulation Method for SC Turbines," a detailed design of the turbine used for SCPPS will be presented, accompanied by a mathematical model and simulation method of the SCPPS coupling the moving part pressure-based wind turbine. In chapter "Energy Storage of Solar Chimney," a detailed discussion on the energy storage characteristic of SCPPS will be presented. In this chapter, analysis of different materials and layout of energy storage will be shown. In chapter "The Influence of Ambient Crosswind on the Performance of Solar Updraft Power Plant System," we will introduce the effect of ambient crosswind on the performance of SCPPS. In chapter "Experimental investigation of a solar chimney prototype," an experimental investigation of an SCPPS setup will be briefly introduced. In chapter "Research Prospects," the future research development will be discussed.

This book is supported by the Key Research Program of the Chinese Ministry of Education (Grant No. 104127), the National Natural Science Foundation of China (Grant No. 51106060), China Postdoctoral Science Foundation Fourth Special Funded Project (Grant No. 201104460), the Natural Science Foundation of Hubei Province, China (Grant No. 2012FFB02214), Scientific Research Foundation of WUT (Grant No. 40120237), and the fundamental research funds for the Central Universities (WUT Grant No. 2016IVA029). Additionally, several contributors have helped the authors to create this book.

A large part of the work of this book was performed when I worked in Huazhong University of Science and Technology and the rest was accomplished in Wuhan University of Technology. The contributors of this book include my two supervisors, Professor Wei Liu from School of Energy and Power Engineering, Huazhong University of Science and Technology (HUST) and Professor Yuan Pan from School of Electrical and Electronic Engineering, HUST. In addition, the contributors also include Professors Shuhong Huang, Suyi Huang, Guoliang Xu, and Tianhua Wu, Drs Renaud K. de Richter, Xuemin Li, Xiaoming Huang, Aiwu Fan, and Kun Yang, my students Dr Jun Liu, Mr Xiaoyang Shi, Yong Zheng, Xinjiang Wang, Fanlong Meng, Yongjia Wu, Wenqing Shen, Lixian Wang, Tao Fang, Zhou Zhou, Jinle Gui, Chao Liu, Tao Pan, and Keyuan Peng, Ms Xiangfei Yu, Cheng Zhou, Yue Qiu, and so on. Drs Xiaohu Liu, Hui Liu, Dongyuan Shi, Xinchun Lin, and Xinping Zhou are greatly appreciated for their kind suggestions on the work of this book.

Tingzhen Ming

January 20, 2016

Introduction

1

Tingzhen Ming[1,2], Wei Liu[2], Yongjia Wu[2], Jinle Gui[2], Keyuan Peng[2]
and Tao Pan[2]

[1]*School of Civil Engineering and Architecture, Wuhan University of Technology,
Wuhan, P.R. China* [2]*School of Energy and Power Engineering, Huazhong
University of Science and Technology, Wuhan, P.R. China*

CHAPTER OUTLINE

1.1 ENERGY BACKGROUND

1.1.1 THE ENERGY ISSUE AND THE STATUS QUO

Energy is the lifeblood of the national economy and closely related to the living environments of human beings. Since the global energy crisis occurred in the 1970s, the depletion of fossil energy resources has caused economic recession in many developed countries, which affects the sustainable development of the national economy and social stability directly. The main cause of the wars and

Table 1.1 Oil Production and Consumption in 2013 (Millions of Barrels Per Day) [1]

Rankings	Countries	Production	Countries	Consumption
1	United States	12.343	United States	18.961
2	Saudi Arabia	11.702	China	10.303
3	Russia	10.764	Japan	4.531
4	China	4.501	Russia	3.515
5	Canada	4.073	India	3.509
6	United Arab Emirates	3.441	Brazil	2.998
7	Iran	3.192	Saudi Arabia	2.968
8	Iraq	3.058	Canada	2.431
9	Kuwait	2.812	Germany	2.403
10	Brazil	2.694	South Korea	2.324

terrorist threats all around the world in recent decades should be attributed to the exploitative ways the developed countries adopted in controlling fossil energy resources. There is no doubt that, the overexploitation of fossil fuels has caused worldwide environmental pollution, global warming, melting glaciers in the Arctic region, human diseases increasing, and deterioration in the ecological environment. While developed countries sustain a high standard of living quality and economic development rate by relying on an excessive use of fossil fuels, the developing countries with the vast majority of the population of the world still cannot afford the necessary cost of traditional energy resources. In addition, they are not able to compete with the developed countries in the fight for fossil energy resources. The people in these countries are still living in poverty, backwardness, and lacking electricity and clean water, which in turn leads to the population surge and aggravates poverty. The protection of the Earth's ecological environment, the sustainable progress of civilization, and the stability of the international community mainly depend on government regulation and control around the world, the environment protection sense of the public, the saving and clean use of fossil energy resources, the generalization and application of renewable energy, the well-organized control on the population growth in developing countries, and so on.

Consumption and production for all fuels except nuclear power have increased remarkably during the recent decades. For each of the fossil fuels, global consumption rose more rapidly than production. Table 1.1 shows the oil production and consumptions of the top 10 world countries in 2013 [1]. It indicates that the United States and China utilize a very large percentage of the world's oil consumption.

Due to the increased consumption of fossil energy, the global CO_2 emissions have grown significantly, with China and the United States ranking No. 1 and 2 since 2009 [2]. According to the Department of Energy's (DOE) Energy Information Administration's (EIA) forecasts for emissions from energy use until 2030 (Fig. 1.1), this trend will last if we do not intervene.

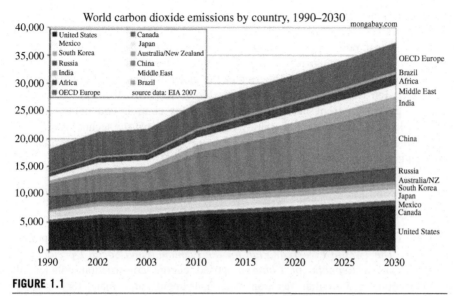

FIGURE 1.1

Past and projected CO_2 emissions for countries, 1990–2030 [3].

China has become one of the largest countries in energy production and consumption. The huge total reserves but relatively low per capita possession, uneven distribution, and low utilization efficiency are four key features of energy production and consumption in China. The total energy production in China is only less than the United States and Russia, ranking No. 3 in the world. The necessary energy use accounts for 10% of the world, ranking only second to the United States. The annual average economic growth rate for China was 9.7% while the growth rate of energy consumption was only 4.6% in the past 20 years. Since 1980, the total energy consumption for China has grown about 5% per year, which is nearly three times the world's average growth rate. As China's demand for energy keeps increasing, there is an enormous gap between China's energy reserves and future development needs. The total amount of the energy gap was over 100 million tons of standard coal in 2003. It is predicted that the total energy gap will be about 250 million tons of standard coal in 2030, and it will reach about 460 million tons of standard coal in 2050. With the enlargement of the energy gap year by year, China's dependence on energy imports will expand gradually, which is the primary concern for our energy security in the future. In addition, since China's main energy source is coal, the conflict between economic development and environment pollution should be much more severe. Coal combustion produces a lot of CO_2, SO_2, NO_x, and many other harmful gases, which are blamed for the greenhouse effect and acid rain. The emissions of trace elements and particulate matter in the coal combustion processes also threat human health.

The world's energy reserve is not good news. Besides coal, oil, and natural gas reserves could be mined for less than 100 years. Despite the rapid development of nuclear power plant, uranium will be available for less than 100 years since the world's uranium reserve is quite low. As for China, the recovery time will be only about 50 years. Furthermore, the general public have doubts on the nuclear fission technology and technology blockade. Some technical problems, such as protection from nuclear radiation, reactor control, and nuclear waste disposal [4−11], add to the uncertainty about the future of the technology. Thus, large-scale construction of nuclear power plants worldwide is not a permanent solution to the energy problem.

1.1.2 CHINA'S ENERGY POLICY AND PROSPECT

In view of the energy, environmental, economic, and sustainable development problems, *The Twelfth Five-Year Plan for the Economic and Social Development of the People's Republic of China* proposed energy construction policies of "implementing preferential taxation and investment, and mandatory market share policies to encourage the production and consumption of renewables and increase its proportion in primary energy consumption," "Positively developing and utilizing solar, geothermal and ocean energy," "Strengthening the exploitation and application of air water resources, solar, wind and other energies." At the same time, the plan introduced the environmental protection policies of "the essential point of ecological protection transiting from post-treatment to protection in advance." In addition, a hard target of a 4% energy saving during "Twelfth Five" is proposed. According to the development goals determined by *The New and Renewable Energy Industry Development Plans and Key Points in 2000−2015*, China's new and renewable energy production would reach 43 million tons of standard coal by 2015, which is about 2% of the total energy consumption. China would cut more than 30 million tons of greenhouse gases emissions and more than 2 million tons of sulfur dioxide emissions. The visible effect would be indicated by a reduction in atmospheric pollution and an improvement of the atmospheric environment quality. Nearly 500 thousand jobs would be provided, and more than 5 million farmer families (more than 25 million people) living in remote areas would be alleviated from the lack of water and electric. So, exploitation and utilization of renewable energy is a strategic direction for China's energy development. In recent years, the exploitation and use of new and renewable energies has developed rapidly throughout the world, with technologies being matured gradually, and economically viable products and equipment being put on the market quickly. It is expected that the development and utilization of the new and renewable technologies and market share of these technologies will breakthrough dramatically in the coming decades. The new and renewable technologies have broad prospects for development (Fig. 1.2).

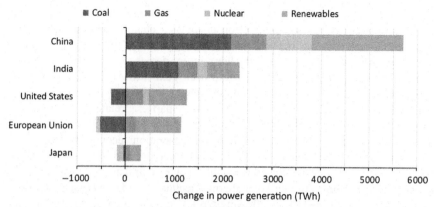

The need for electricity in emerging economies drives a 70% increase in worldwide demand, with renewables accounting for half of new global capacity

FIGURE 1.2

Global change in power generation, 2010–2035 [12].

In the development of energy technologies in China, a challenging problem is how to develop and utilize solar energy on a large scale and to eliminate the dilemma of a shortage of conventional energies gradually. Solar energy is clean and inexhaustible. China is rich in solar energy resource. The regions with a total solar radiation of more than 5020 MJ/m^2 annually and a sunshine time of more than 2200 h cover two-thirds of the total territory of China, which are good conditions for the development and application of solar energy technologies. Especially in China's western regions, where the solar energy resource is very rich, the population is small, the ecological environment is fragile, and the desertification situation is very severe. The acute shortage of water and electricity is one of the most important factors that restricted the economic development in the regions. Developing solar power technologies can not only improve the economic and ecological environment in the western region, but can also effectively utilize waste land. The action is in line with the national energy and environmental coordination and management policies. Solar power technologies surely have very broad application prospects.

1.1.3 SOLAR POWER GENERATING TECHNOLOGIES AND THE STATUS QUO

The techniques that utilize working fluids or receiving devices to convert solar radiation into electricity eventually by some way are referred to as solar power generation techniques. Currently, there are two main kinds of solar energy generating technologies. The first kind is solar thermal power generation technologies, that is, they convert solar radiation into heat first, and this is followed by a particular power generation process to change thermal energy into electrical energy, such as thermoelectric power generation utilizing semiconductor or metallic materials [13−19], thermionic power generation in vacuum devices [20−24], alkali metal thermal power generation [25−32], and MHD power generation [33−38], and so on. The characteristics of these technologies are power generation devices with no moving parts, a relative small electric generation capacity, and they are still in the primary experimental phase for many technologies. Currently, thermal power generation technologies are issues that are the most interested in, researched most deeply, and the most promising worldwide [39−45]. Technologies, including solar central power tower technology [46−50], parabolic trough solar thermal technology [51−57], and dish solar thermal technology [58−64], use flowing work mediums to convert the solar radiation into thermal energy, and then drive the generator by heat engine to convert the heat energy of the medium into electricity. The basic equipment compositions of these technologies are similar to conventional power generation equipment [65]. Other solar power generating technologies converting solar energy into electrical energy directly are light induction power generation [66], photochemical power generation [67], and biological power generation [68,69]. The photovoltaic power generation technology [70−77], which transforms the solar radiant energy into electrical energy through the solar battery, is successfully commercialized. A brief introduction of several major solar thermal power generation technologies follows.

1.1.3.1 *The solar central power tower system [47,78−85]*

The solar central power tower system generates electric power from sunlight by concentrating solar radiation on a tower-mounted heat exchanger (receiver). The system uses hundreds to thousands of sun-tracking mirrors called heliostats to reflect the incident solar radiation onto the receiver. These kinds of solar power plants are best suited for utility-scale applications in the 30- to 400-MWe ranges. In a molten-salt solar power tower, liquid salt at 290°C is pumped from a "cold" storage tank through the receiver where it is heated to 565°C and then on to a "hot" tank for storage. When power is needed from the plant, hot salt is pumped to a steam generating system that produces superheated steam for a conventional Rankine-cycle turbine/generator system. From the steam generator, the salt is returned to the cold tank where it is stored and eventually reheated in the receiver.

In 1981, the United States successfully built a pilot solar-thermal project, that is, the Solar One solar central power tower plant, in the Mojave Desert just east of Barstow, California with 10 MW installed capacity. It was the world's first test prototype of a large-scale thermal solar power tower plant. Solar One was designed by the DOE, Southern California Edison, LA Department of Water and Power, and California Energy Commission. The energy collection method of Solar One was based on concentrating the solar energy onto a common focal point to produce heat to drive a steam turbine generator. It had hundreds of large mirror assemblies (heliostats) which tracked the sun, reflecting the solar energy onto the tower erecting in the center of the mirror area where a black receiver absorbed the heat. High-temperature heat transfer fluid was used to carry the energy to a boiler on the ground where the steam was used to spin a series of turbines, much like a traditional power plant. The primary difference between the solar central power tower system and the coal-fired power plants is that the former uses solar receiver to collect energy which has the same function as the boiler.

Solar One was converted into Solar Two (Fig. 1.3) in 1995 by adding a second ring of 108 larger 95-m^2 heliostats around the existing Solar One. Solar Two was put into test operation in January, 1996 and was decommissioned in 1999. The 1926 heliostats occupied a total area of 82750 m^2. As a result, Solar Two had the ability to produce 10 MW to power an estimated 7500 homes. In addition, Solar Two used molten salt, a combination of 60% sodium nitrate and 40% potassium nitrate, as an energy storage medium instead of oil or water as with Solar One, which helped in energy storage during brief interruptions in sunlight due to clouds. The molten salt also allowed the energy to be stored in large tanks for future use such as at night time—Solar Two had sufficient capacity to continue running for up to 3 h after sunset.

FIGURE 1.3

Solar Two in Mojave Desert of California, USA [86].

Although the tower thermal power generation system starts earlier, it has high system cost, low installed capacity, and the industrialization of the system faces a lot of problems. The primary reason for all the problems is the design of heliostat system. Nowadays, the typical heliostat of tower thermal power generation has two characteristics. First, the typical heliostats almost adopt the ordinary spherical or flat reflective surface. Second, tracking angles for the heliostats all use the traditional azimuth elevation formula. These two features result in the following problems in the tower solar concentrator receiver: (1) The facula focusing on the receiver changes substantially, which causes the concentrated light intensity to fluctuate significantly. The ordinary spherical or plane reflector cannot overcome aberration caused by the motion of the sun, which leads to the heat conversion efficiency of the tower system being only 60%. Although methods to design heliostats spherical surfaces with different curvature radii were developed to reduce the sunspot size, the optical design complexity is significantly increased, which leads to the rise of the manufacturing costs. (2) Many heliostats are built around the center tower and occupy a lot of land. To concentrate the sun on the top of the center tower efficiently, each heliostat cannot share the light of others and then the distance between near heliostat rows will increase along with their positions in the center tower. Therefore, the area of the tower thermal power generation system will increase exponentially along with the increase of power generation capacity. (3) Each heliostat requires a separate two-dimensional control, and the control system is quite complex. In a tower system, each heliostat has a different position relative to the center of the column. Therefore, each heliostat tracking should be a separate two-dimensional control. The control of each heliostat is different, which makes the control system complicated and unreliable, in particular for the installation of optical alignments. (4) In order to alleviate the cosine effect of heliostats, the center tower should be built very high. The center tower of the Solar Two (10 MW) thermal power is up to over 100 m. This leads to not only the dramatically increasing cost of the thermal power generation system, but also the awful adaption ability of the system to the harsh windy weather.

1.1.3.2 The parabolic trough solar thermal system [54,56,57,87–95]

The parabolic trough solar thermal electricity generation system (TSTEGS) collects high-temperature thermal energy by cascading a large number of parabolic trough concentrating collectors. Then a heat medium in the pipelines produces super-heated steam and drives a steam turbine generator to generate electricity. LUZ Solar Thermal Electricity Generation International Co., Ltd built 9 plants successively, with a total capacity of 354 MW, in the Mojave Desert region of Barstow, California, from the middle 1980s to the early 1990s (Fig. 1.4). The company was trying to reduce the generation cost from 24 cents/kWh to 5 cents/kWh, but, unfortunately, the company's bankruptcy led to the stop of the study plan.

FIGURE 1.4

Part of the 354 MW SEGS solar complex in northern San Bernardino County, California [96].

TSTEGS replaced the point focusing with line focusing, and the focusing pipeline tracked the sun together with the cylindrical parabolic mirrors. TSTEGS solves the problem that the light heat conversion efficiency is not high due to the nonuniform focusing spot, lifting the light heat conversion efficiency to approximately 70%. But the parabolic trough solar thermal system also brings three new problems: (1) The system cannot track the sun with a fixed angle, and the tracker is cumbersome and unwieldy. The solar receiver (focusing pipeline in the core of the tractor) is fixed in the reflector trough and moves together with the reflector, resulting in a relatively bulky system. Meanwhile, the connection section of the heat pipe must be active and this structure is likely to damage the insulation and break down. (2) Its wind resistance ability is poor, and not suitable for working in the wind. Each parabolic trough reflector is a larger body of the lens with the size of 99 m long and 5.7 m wide. The wind resistance is very large. So the existing trough type solar thermal power generation system is commonly used in desert regions with no wind or breeze environment that are significantly different from the windy environments in northeast China. An improvement must made to reinforce the mirror supporting structure to strengthen the wind resistance performance of the parabolic trough reflector before it can be successfully applied in China. But, without doubt, it will inevitably lead to a sharp rise in initial investment cost and thermal power generation cost. (3) The receiver of the trough system is long, and the cooling area is large. The solar receiver of the TSTEGS is a very long heat pipeline. Though many new light-absorbing technologies have been developed, its heat dissipating area is much larger than its light-absorbing area. When compared with other typical concentrator systems, the TSTEGS is relative bulky.

1.1.3.3 The Dish-Stirling solar power plant system [64,97–103]

The Dish-Stirling solar power plant system (Fig. 1.5) takes advantage of rotating parabolic mirrors to collect solar radiation energy to drive electricity generators. As with the trough-type system, parabolic dish solar receivers are not fixed. Dish mirrors track the movement of the sun to eliminate the large energy loss of cosine effect in the solar central power tower system, so the thermal conversion efficiency is significantly improved. However, unlike the trough type system, the trough receiver focuses the sun radiation on the focal line of the parabolic surface while the dish receiver focuses solar radiation on the focus of the parabolic dish. This significantly improves the temperature performance of the dish-type system. In 1983, the Jet Propulsion Laboratory in California built a Dish-Stirling solar thermal power system with a concentrator diameter of 11 m, a maximum power output of 24.6 kW, and a conversion efficiency of 29%. In 1992, a German company developed a Dish-Stirling solar thermal power system with an electricity generation capacity of 9 kW, with a peak efficiency of 20%.

The challenges of the Dish-Stirling systems are: (1) the cost of the system is the highest among the three systems. The initial investment cost of the dish-type thermal electricity generation system is as high as 47000−64000 yuan/kW; (2) although concentrated the ratio of the dish system is very high, up to 2000°C, under the present thermal power generation technology situation, there is no need of such a high temperature, because the too high temperature will damage the system. So, the receiver of the dish-type system is not at the focus point generally.

FIGURE 1.5

A Dish-Stirling solar thermal power generating system [104].

It is always set in a low-temperature region in accordance with the performance requirements. The advantages of a high light concentration is not displayed; (3) hot melt salt heat storage technology cost is high and dangerous.

1.1.3.4 The solar photovoltaic power generation system [70,72,73,75–77,105]

The solar photovoltaic power generation system is mainly composed of photovoltaic cells, a solar controller, a storage battery, and so on. When the output power is required to be 220 V or 110 V, an inverter is necessary. The development of solar photovoltaic cells has been very rapid. It has been used in various industry and agriculture fields.

The use of solar energy is growing strongly around the world, in part due to the rapidly declining costs of solar panel manufacturing. In 2011, the global total solar PV installed capacity was 68850 MW, and the corresponding global total solar PV actual generation was 52878 GWh [106].

But the photovoltaic system has the following problems: (1) Chinese domestic solar photovoltaic module manufacturers are small with backward infrastructure and production equipment. Product overall quality is lower than foreign goods. The quality of the local packaging material has not met the expected requirement, and parts of the packaging materials should be imported, which increases the cost. The photovoltaic module production cost is 20% higher than abroad. (2) There is still no uniform photovoltaic power station design and construction specifications at home or elsewhere. So the quality of the system cannot be guaranteed, and the further promotion of photovoltaic power generation will be effected significantly. (3) The amount of money for research and development is limited. The bank loan interest rate is too high. To develop the photovoltaic project, the problem of long-term financing channels needs to be solved. (4) In China, there are no executive incentives to encourage the development of photovoltaic power generation, such as tax, subsidy, photovoltaic grid, reasonable electricity price policy, etc. (5) The lack of infrastructure, small local production capacity, poor sale and service, consumer unawareness, and no new products put into markets will significantly limit the development of solar photovoltaic system.

1.2 SOLAR CHIMNEY POWER PLANT SYSTEM

1.2.1 THE APPEARANCE OF A SOLAR CHIMNEY POWER PLANT SYSTEM

In order to achieve coordinated development of China's energy and environment, conventional energy generation technologies are not enough. The use of the existing new energy technologies can not completely solve the problem, so we must continue to seek some alternative ways to solve the problems. We should solve not only the problem of energy exploitation, but also the problem of sound ecological environment protection. Only in this way, can we ensure the sustainable development of the Chinese economy. In terms of power generation technologies, the costs of developing the solar photovoltaic and thermal power generating technologies, including tower, trough, and Dish-Stirling systems, are not able to compete with coal-fired power stations. Thereby, constructing large-scale power plants in MW scale and above are not feasible.

The solar chimney power plant system (SCPPS), sometimes called solar updraft tower (SUT), is a renewable energy power plant for generating electricity from solar radiation [107]. The system is composed of four parts: chimney, collector, energy storage layer, and power conversion units (PCU) (Fig. 1.6). The main objective of the collector is to collect solar radiation to heat up the air inside. As the air density inside the system is less than that of the environment at the same height, natural convection affected by buoyancy, which acts as the driving force, comes into existence. Due to the existence of the chimney which is erected in the middle of the collector, the cumulative buoyancy results in a large pressure difference between the system and the environment. The heated air then rises up into the chimney with great speed. If an axis-based turbine is placed at the bottom of the chimney or near the outlet of the collector where there is a large pressure drop, the potential and heat energy of the air can be converted into kinetic energy and ultimately into electric energy.

The idea of utilizing the solar chimney technology in power generation was originally put forward by Isidoro Cabanyes in 1902 [108]. In 1970s, Professor Schlaich restated this idea in some conferences [109] and built an SCPP prototype in Manzanares, Ciudad Real, 150 km south of Madrid, Spain [110−113]. The height of power station chimney was 194.6 m, the diameter of the chimney was

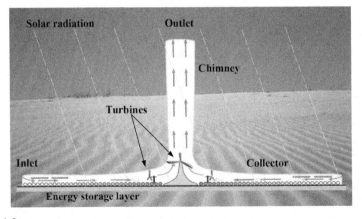

FIGURE 1.6

Schematic of a solar chimney power plant system.

10.8 m, and the collection radius was 122 m. The canopy of the collector sloped from the inlet to the center, with the height increasing from 2 to 6 m linearly. The plant design capacity was 50 kW while the actual average power output was 36 kW. During the 7 years continuously running period, the average chimney exit air velocity was 15 m/s under the condition of no-load. The full load average speed was 9 m/s with a low operation cost. Researchers spent 9 years continuously improving the design. During the 7-year service time, the power plant's running time exceeded the expected 95% [114,115]. The power plant operated for approximately 7 years. The tower's guy-wires were not protected against corrosion and failed due to rust and storm winds. The tower blew over and was decommissioned in 1989 [116]. From then on, the research on SCPP attracted worldwide attention.

1.2.2 ADVANTAGES OF SCPPS

The SCPPS is a not new type of solar power generation system as it was verified several decades ago. However, when compared with the traditional power generating methods, it has the following advantages: easier to design, more convenient to draw materials, lower cost of power generation, higher operational reliability, fewer running components, more convenient maintenance and overhaul, lower maintenance expense, no environmental contamination, continuous stable running, and longer operational lifespan. In addition, it can partly meet the electricity demand in developing countries and regions where traditional power resources are limited. In more detail, its primary characteristics are manifested in the following aspects:

1.2.2.1 Large scale renewable energy collection

Low energy flux density is the shared characteristic of almost all renewable energy sources; and how to collect renewable energy at a large scale becomes a challenge to all renewable energy technologies. Low energy flux density and large sunshine fluctuation are the fundamental characteristics of solar radiation, and they are also an insuperable barrier for humans in the large-scale exploitation and utilization of solar energy. However, by building SCPPS with an output power of up to 100 MW, the diameter of the collector area will be several kilometers, and it is easy to collect a large amount of low-density solar energy at a low cost.

1.2.2.2 Energy storage with low cost

Considering the fluctuation and intermittence of renewable energy resources, energy storage systems play a significant role in renewable energy technologies, especially the renewable energy power plants. There are many different energy storage technologies [55,117−121], but their prices are very high and the corresponding storage capacities can last only a few hours. The underground materials of the collector serve as a large energy storage system. The solar radiation hits the ground surface through the transparent collector canopy and heats the ground materials. The solar energy is thus stored and releases continuously to the air within the collector canopy which keeps the system operating uninterrupted even on rainy days. The energy storage materials under the collector can be soil, sand, stone, and hermetic water. These materials are very cheap and easy to get in any local area. In the daytime, the energy storage materials can absorb and store energy under direct sunlight. At night, with the radiation condition changes, the heat is released to sustain a stable power output.

1.2.2.3 Air as working fluid

For the SCPPS, air is the only working fluid without phase change. The solar radiation energy resource is abundant in West China. Most areas there suffer from severe water shortage and, therefore, it is not feasible to construct large-scale power systems using water as a working fluid and/or water as a cooling fluid [51,97,122]. Since the working fluid of SCPPS is air, there is no phase change, no water demand, no working mediums, or cooling equipment device in the operation process, which significantly reduces the complexity of the system.

1.2.2.4 Technical feasibility

The SCPPS are composed of a collector, a turbine, a chimney, and an energy storage layer. All these technologies are available and widely used. A key novelty of SCPPS is that it combines these simple technologies to generate electricity without any difficulty. The moving parts are only the turbines and generator, and as a result, the operation and maintenance cost is very low. In addition, the construction, operation, and maintenance of the SCPPS will provide large numbers of jobs for the local people, which is a unique advantage of constructing this type of power station in developing countries.

1.2.2.5 Environmental remediation

Generally, the SCPPS is built in deserts and useless land, and this technology can thus take advantage of the western desert land without immigration needed. Compared with the large hydropower station, SCPPS does not significantly change the local environment and climate. Further, this technology can improve the environment and ecosystem of western China by alleviating the water and electricity shortage. SCPPS can partly take the place of fossil-fired power plants and reduce CO_2, SO_2, and NO_x emissions. Meanwhile, it can be employed to improve the local environment by taking advantage of the its greenhouse effect.

1.2.2.6 Competitive investment and operation costs

The system is simply designed and the building materials, such as glass, cement, steel, are available in the local area. The construction cost of the power station is acceptable, with initial investment being expected to be similar to the cost of building a hydropower plant with the same installed capacity. Land covered by the collector can be reused to plant flowers, grass, and vegetables, which will also increase the air humidity and thus improve the air quality of the ambience around the SCPPS.

1.2.3 WEAKNESSES OF SCPPS

However, an inevitable problem is that the overall efficiency of SCPPS is relatively small. The overall efficiency of the SCPPS is influenced by the greenhouse efficiency of the collector, the updraft efficiency of the chimney, the thermal to mechanical efficiency of the turbines, and the mechanical to electrical efficiency of the generators. The chimney plays an important role in increasing the overall efficiency of the SCPPS; the higher the chimney, the higher the overall efficiency. The collector is a colossal solar energy collection system; the larger the collector diameter, the larger the system output power and energy being stored. So this kind of power stations is not suitable for those areas near metropoles where the land is very expensive.

Considering the commercial application of an SCPPS with an output power up to 100 MW, the collector diameter should be several kilometers, which will cause difficulty in cleaning the collector canopy; the chimney should be about 1000 m, which will be a challenge for construction as there are no buildings in the world this high presently.

1.3 RESEARCH PROGRESS

1.3.1 EXPERIMENTS AND PROTOTYPES

The concept of the solar chimney has been proposed for over 100 years. In 1903, Isidoro Cabanyes, a colonel in the Spanish army, was originally proposed a solar chimney power plant in the magazine La energía eléctrica (Fig. 1.7) [123]. In 1926 Professor Engineer Bernard Dubos proposed to the French Academy of

FIGURE 1.7

The first concept of solar chimney in 1903.

FIGURE 1.8

Solar chimney in the Moroccan desert envisioned by Bernard Dubos.

Sciences the construction of a Solar Aero-Electric Power Plant in North Africa with its solar chimney on the slope of a large mountain (Fig. 1.8) [123]. In 1931, a German researcher, Hanns Günther advanced a solar chimney power generating technology [124], this technology simply used the stack effect and greenhouse effect to drive a turbine. Several decades later, this technology did not draw the researchers' attention until Professor Jörg Schlaich, a Stuttgart University, put forward the same idea again in a conference in 1974 [109].

In 1976, Professor Schlaich thought of electricity generation by using the updraft effect, when he was studying the use of circulating water cooling towers to cool the nuclear plant up to 800 m high. Four years later, Professor Schlaich proposed the concept of solar chimney power generation at the annual German Energy Conversion and Utilization Meeting [109]. Professor Schlaich's thought immediately aroused a heated discussion of the participants and the former West German government attached great importance to it. In 1981, the former West German government research department, cooperating with the Spanish Electricity Company, provided 15 million Deutsche mark to build the world's first solar chimney power plant prototype in Manzanares, 150 km south of Madrid [110−112] (Fig. 1.9). The system is composed of a chimney 200 m in height and 10 m in diameter, a greenhouse collector with a radius of 122 m and a height of 2−6 m from the periphery to the center (Fig. 1.10), and a controllable wind turbine with a rated power of 50 kW (Fig. 1.11). The chimney is erected in the center of the collector and is fixed by cables (Fig. 1.12). Based on this prototype,

FIGURE 1.9

The prototype of solar chimney power plant in Manzanares, Spain.

FIGURE 1.10

The collector and the plants.

Haaf [114,115] and Lautenschlager [125] carried out detailed research on the experiment system design standard, energy balance, cost analysis, the experimental results, and the design strategies for different scale solar thermal power generation system.

FIGURE 1.11

Turbine at the chimney bottom.

FIGURE 1.12

Chimney, collector, and the cables.

Since then, researchers from many countries have built solar thermal power generation systems of different sizes and types. In 1983, Krisst [126] made a "backyard" SCPPS experiment device in West Hartford, Connecticut in the United States. The height of the instrument stack was 10 m, and the diameter of the

FIGURE 1.13

Solar chimney power plant designed by Professor Sherif's group.

collector was 6 m with the output power up to 10 W. In 1985, Kulunk [127] built a miniature SCPPS demonstration unit in Turkey Izmit. The device stack was only 2-m high, the chimney diameter was 70 mm, and collector area was 9 m² with an average power output of 0.14 W. Kulunk reported that the turbine shaft power was 0.45 W, the electricity generating efficiency was 31%, the chimney inlet and outlet temperature difference was 4 K, and the differential pressure was 200 Pa.

In the 1990s, the research group of Professor Sherif [128−133] built three different types of small-scale SCPP model at the University of Florida in Gainesville, United States (Fig. 1.13) and conducted a series of experiments to verify the operating mechanism and energy storage performance. The test units made significant improvements in the following two aspects: (1) The outside edge of the collector was made of the slope type extension; (2) The heat storage medium was introduced as a heat absorber. The extension of the collector area helped to raise the air temperature, and the heat storage medium not only increased average air temperature, but also increased average the mass flow rate. These two approaches increased the output power of the system. The researchers conducted detailed experiments on the system. They obtained the impacts of the solar radiation, the chimney height, collector radius and heat storage layer on stream temperature, speed, and the output power of the system. The cost of investment for different scale SCPPs was predicted, and the results were encouraging, which indicated that large-scale solar thermal power generation system was feasible.

FIGURE 1.14

Akbarzadeh et al.'s test model combining a solar pond [134].

A 3000-m^2 solar pond was constructed in Pyramid Hill as part of the facilities of the Pyramid Salt Company in northern Victoria (Fig. 1.14) [134]. The pond was commissioned in June 2001 and reached a temperature of 55°C in February 2002. It has the capability of producing a minimum of 50 kW heat on a continuous basis. Later, the concept of combining a salinity gradient solar pond with a chimney to produce power in salt affected areas was examined [135]. Firstly the causes of salinity in salt affected areas of Northern Victoria, Australia are discussed. Existing salinity mitigation schemes are introduced and the integration of solar ponds with those schemes is discussed. Later it is shown how a solar pond can be combined with a chimney incorporating an air turbine for the production of power. Following the introduction of this concept the preliminary design is presented for a demonstration power plant incorporating a solar pond of area 6 hectares and depth 3 m with a 200-m tall chimney of 10-m diameter. The performance, including output power and efficiency of the proposed plant operating in northern Victoria is analyzed and the results are discussed.

Zhou et al. [136] designed and built a pilot solar chimney power setup in Wuhan China, in December, 2002 in order to investigate the temperature field as well as to examine the effect of time of day on the temperature field (Fig. 1.15). The collector roof was made of glass with a thickness of 4.8 mm, and chimney was made of PVC [137]. In order to avoid diffusing heat through the collector canopy by convective heat transfer, a heat insulator was used to pack the steel-frame

FIGURE 1.15

Zhou's SCPPS test model [136].

structure of the collector. In addition, several pipes with a diameter of 6 cm containing water were paved on the surface under the collector as the energy storage system to store solar radiation. Above the water pipes, one centimeter thick composite layer bed with asphalt and black gravel was applied as the top layer to absorb solar radiation, and then heat was transferred from the top layer to water pipes. At the entrance of the chimney, a multiple-blade turbine was designed and installed to test the power output of the pilot SCPPS. A shield was installed a few centimeters above the chimney top to avoid ambient crosswind entering down into the chimney [136].

In 2007, Koyun et al. [138] built a pilot solar chimney on the campus of Suleyman Demirel University-RACRER (Research and Application Center for Renewable Energy Resources), in Isparta, Turkey (Fig. 1.16). The prototype had a chimney with a height of 15 m and diameter of 1.19 m and a glass covered collector 16 m in diameter. The greenhouse area was about 200 m^2, and the inlet area in the periphery of the collector was 31.15 m^2. Both experimental studies and theoretical modeling have been performed to test the design.

In 2008, Ferreria et al. [139] built a physical prototype in Belo Horizonte, Brazil for food drying. In this pilot prototype, a tower 12.3 m in height was constructed with sheets of wood and covered by fiberglass at a diameter of 1.0 m. The cover with a diameter of 25 m was made of a plastic thermo-diffuser film. The cover was set 0.5 m above the ground level, using a metallic structure. The absorber ground was built in concrete and painted in black opaque (Fig. 1.17). Ferreria et al. [139,140] and Maia et al. [141−144] presented a series of

FIGURE 1.16

Koyun et al.'s SCPPS test model [138].

FIGURE 1.17

Ferreria et al.'s SCPPS test model for food drying [139].

theoretical analyses and experiments on the system for food drying. The results indicated that the height and diameter of the tower are the most important physical variables for solar chimney design.

In 2011, Kasaeian et al. reported a SCPPS prototype at the University of Zanjan, Iran [145]. The construction of the chimney required the use of 10 iron cylindrical

FIGURE 1.18

Kasaeian's SCPPS test model [145].

barrels with a height of 1.6 m and a diameter of 0.2 m. These barrels were butt-welded. To maintain the chimney, two collars were used at 6.5 and 12.5 m above the ground. For each of them, steel cables were used. A turbine was installed at a height not exceeding 0.55 m. the platform of the turbine was installed downwardly. The blades of the turbine were made of aluminum. The diameter of the turbine (vertical axis) chosen to prevent air leakage was equal to 0.36 m (Fig. 1.18).

Without any shadow of a doubt, energy production based on renewable energies is one of most fundamental methods for energy generation for the near future. After designing and making a solar chimney pilot power plant with a 10-m collector diameter and a 12-m chimney height, the temperatures and air velocities were measured. The temperature and velocity readings were carried out for some specified locations within the collector and chimney with varying parameters on different days. Because of the greenhouse effect happening under the collector, the temperature difference between the collector exit and the ambient reached 25°C, which caused a creation of air flow from the collector to the chimney. The air inversion at the bottom of the chimney was observed after sunrise, on both cold and hot days. The air inversion appeared with increasing solar radiation from a minimum point and after a while, it was broken by the collector

FIGURE 1.19

SSUPP prototype in Damascus University, Syria [146].

warm-up. After the inversion breaking, there would be a steady air flow inside the chimney. The maximum air velocity of 3 m/s was recorded inside the chimney, while the collector entrance velocity was zero.

Kalash et al. [146,147] reported a pilot sloped solar updraft power plant system (SSCPPS) built in the south campus at Damascus University, Syria (Fig. 1.19). The sloped solar collector has a triangular shape and is tilted at 35° toward the south with an approximate area of 12.5 m². The chimney height is 9 m and the diameter is 0.31 m. Eighteen temperature sensors were installed inside the sloped solar collector to measure glass, air, and absorption layer temperatures at different points along the collector. Practical data were recorded every 10 minutes for a total of 40 consecutive days to investigate the temperature changes in the sloped collector. Experimental results indicated that, although measurements were taken during the winter season, the air temperature increased to reach a maximum value of 19°C, which generates an updraft velocity in the chimney with a maximum value of 2.9 m/s. As a result, the solar radiation and the ambient temperature have a direct impact on varying the air temperature between the collector outlet and the ambient. It was also noticed that the absorption layer emits most of its stored thermal energy, which is absorbed from the morning to noon, back to the collector air in a short period of time when solar radiation starts to decrease in the afternoon.

In 2010, Wei et al. [148] built a new solar updraft power plant combined with a wind energy system to generate electricity in a desert of Wuhai, Inner Mongolia of China (Fig. 1.20). The chimney height was 53 m and the diameter was 10 m. The collector canopy covered an area of 5300 m². It was designed based on the solar updraft power by adding some controllable air intakes. There are several air intakes facing north in order to absorb the natural wind, and this can take advantage of the natural wind energy to generate electricity. Thereby, the solar updraft

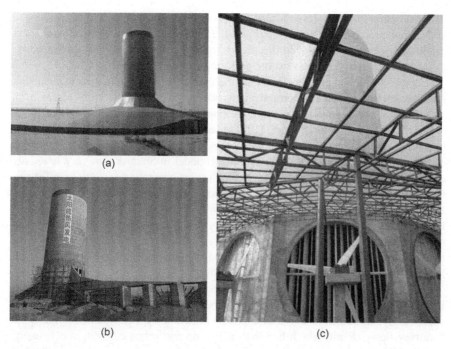

(a)

(b)   (c)

FIGURE 1.20

Wei's solar heated wind updraft tower power. (a) System from the back view. (b) System and the air intakes from the front view. (c) Wind turbine around the chimney and the canopy [148].

power plant combined with wind energy can be applied on both solar radiation and wind energy for power generation. It was reported that the output power of the system was 200 kW connected to the grid in October 2010.

1.3.2 THEORY RESEARCH

1.3.2.1 The thermodynamic theory for the circulation system

In 1980, Schlaich [109] gave the first preliminary description of the thermo-dynamic processes of a solar thermal power generation system in a conference paper. In 2000, Gannon and Backström [149] regarded the working medium flow in the solar thermal power generation system as a standard closed ideal Brayton cycle. The working medium in the whole collector, that is, from the inlet of the collector to the outlet, experienced a constant pressure absorption process. From the bottom of the chimney to the outlet, the working medium underwent an isolation expansion process. The working medium from the vent outlet cooling to the environment temperature was a constant pressure heat release process and the working medium from high altitude down into the collector could be considered as a heat isolation compression process. Process parameters and equations of the above four processes were obtained and the overall efficiency of the system was calculated. Then they analyzed the actual process of the system and the influence of loss in each actual process on the performance of the system.

Energy conversion efficiency of solar thermal power generation system is another important consideration of thermodynamic theory. While in the early research on solar thermal power generation system, Schlaich et al. [109,113,116,150], Haff et al. [114,115] and Lautenschlager et al. [125] raised the mathematical models to calculate the energy conversion efficiency of the system. The calculations showed that the total energy conversion efficiency of the test plant model in Spanish was very low, only about 0.1%. As for large-scale solar thermal power generation system, its effectiveness can reach 1%. In 1985, Louis [151] made an analysis on conversion efficiency from thermal energy to kinetic energy in the greenhouse. He thought, in the greenhouse of the solar thermal power generation system, the infrared radiation and evaporation of water were two significant energy loss ways. By arranging covering material with high transmissivity for short wavelengths and lower transmissivity for long wavelengths, we can actually reduce the loss.

The concentration of rainwater on the roof of the collector was also a factor to influence energy loss, so a method should be developed to make it leave the roof successfully. In addition, the author put forward the hybrid system for the Solar Chimney Power Plant, which had high irrigation and crops cultivation efficiency. In 1987, Mullett [152] established a complete mathematical model to calculate the overall effectiveness of SCPPS with different sizes based on the principle of energy conservation and heat and mass transfer mechanism. He studied the main factors influencing the efficiency of the system, and forecast, calculated, and analyzed the efficiency of SCPPS by researching the Spain model and plants with power generation capacities of 10, 100, and 1000 MW. Predicted calculation results showed that the system overall efficiency was closely related to the size of the SCPPS. The larger the system scale was, the greater the system overall efficiency was. The efficiency range of the large-scale system was 0.96−1.92%.

Gannon and von Backström [149] not only established the thermodynamic cycle model, but also established the power generation efficiency model for SCPPS. Firstly, they analyzed the standard cycle of the ideal gas in the SCPPS and put forward the relationship between the limiting performance, ideal efficiency, and the main variable characters. Then he carried out a detailed analysis of energy losses in each particular part of the system, including friction and local kinetic energy losses in the chimney, heat collector, and turbine. He obtained the function relation between the limit performance of the actual system, the ideal efficiency, and the main variable characters. Finally, based on the Spanish test plant, Gannon carried out an analysis to verify the accuracy of the theoretical model. He also predicted the performance parameters, such as the efficiency and output power of a large-scale power system.

In 2000, Michaud [153] applied the closed thermodynamic cycle for the ideal gas to analyze rising air natural convection process, distinguishing it from the

Brayton gas turbine cycle. He utilized the average temperature method to simulate the heat endothermic process of the heat source and the exothermic process of the cooling source, resulting in a conclusion that air heat conversion efficiency is close to the Carnot cycle efficiency. Analysis results showed that the efficiency did not depend on whether the updraft was continuous, and whether the heat transfer was in a latent or sensible way. Since then, he made an analysis of the efficiency of the system and the availability of the SCPPS. In 2006, Ninic [154] analyzed the thermodynamic cycle of the SCPPS system. When the working medium went through the collector, the air got available energy. Based on this, he established the function relationship between the air available energy and the heat absorbed, air humidity, air pressure, and height. Then he analyzed the influence of dry air and wet air on the performance of SCPPS system. A model to describe the relationship between air available energy and chimney height was established along with the possibility of replacing the chimney with wind turbine.

There are some problems with the thermodynamics theory of SCPPS. On one hand, Gannon's Brayton cycle was based on the total pressure and absolute temperature. At the same time, he took into account the kinetic energy loss in the cyclic graph, which violated the principle that the points on the cycle graph were status points. On the other hand, energy conversion efficiencies, including limiting efficiency, actual efficiency, and energy efficiency, for different SCPPSs did not compare. Energy losses in every part of the system did not analyze.

1.3.2.2 HAG effect of the system

Where did the driving force of solar thermal power generation system come from? Many researchers named the engine as the Helio-Aero-Gravity Effect (HAG Effect) [155−157]. The mechanism of the HAG effect can be explained as follows. When the sun radiation went through the transparent canopy of the collector, the heat storage layer's temperature rose up. Heat released from the heat storage layer warmed the air near the ground surface. Then the density of the air under the canopy decreased. A large buoyancy force in the chimney aroused strong updraft air flow.

Obviously, the factors influencing the HAG effect were solar radiation, air density, and the earth's gravity. Existing research has shown that the system drive force caused by HAG effect can be written as follows [109,113,114,116,152,155−157]:

$$\Delta p = (\rho_0 - \rho)gH_{chim} \tag{1.1}$$

Eq. (1.1) was widely used in SCPPS [150,158−160]. Meanwhile, Mullet [152] thought HAG effect originated from the pressure difference between the chimney bottom and outlet. Researchers who hold the same viewpoint included Lodhi [155,157] and Sathyajith [156], etc. This opinion aroused some misunderstandings. In addition, the description of the driving force based on the mathematical model of HAG effect was a little bit too simple. Eq. (1.1) cannot correctly predict the HAG effect for different geometric structure, environment conditions, and solar radiations on the SC systems. In addition, there were no relevant research reports about the influence of the HAG effect on the system parameters, such as the output power and the energy conversion efficiency. Therefore, it was necessary to make a further analysis and research on the system HAG effect.

1.3.2.3 Heat and mass transfer theory in the system

The most active research work on the SCPPS was concentrated on the heat transfer and flow characteristics. In 1984, Jacobs and Lasier [161] regarded SCPPS as a natural convection system driven by solar energy and made a preliminary theoretical analysis. Lodhi [162] studied the solar energy collection and storage characteristics of SCPPS. Bouchair and Fitzgerald [163] thought that heat storage was a critical influence factor that affected the air temperature and buoyancy force inside the chimney. He analyzed the influence of the sun azimuth angle on the energy collecting efficiency of the collector. The numerical calculation results showed that the solar azimuth angle has a decisive impact on the heat absorption process of the collector.

During 1988−1998, Sherif's group [129,131−133,164,165] at the University of Florida carried out a thorough theoretical analysis for the SCPPS. A set of flow and heat transfer mathematical models were established. Based on the data of experimental SCPPS, Sherif's team had built a set of internal flow mathematical models to describe the heat and mass transfer processes in the heat collector. In these models, they analyzed the internal thermal resistance's influence on the energy loss. By examining the chimney internal flow characteristics, they thought energy conversion followed Betz theory. They also analyzed the influence of different collector designs on the heat and mass transfer characters in the heat collector. The experimental results [131] fit well with the established mathematical model based on the calculation results. Since then, in order to facilitate calculation, Padki et al. [164] developed a very simple mathematical model to forecast heat and mass transfer processes and the power output characteristics of the system. The results showed that the deviations of the simplified models were within 6% [133,164].

Bernardes [166,167] established a whole mathematical model of heat and mass transfer processes in SCPPS, and described its characteristics. The model estimated output power of the system and analyzed the influence of system structure parameters and the external environment temperature on the system output power. Comparing the calculated results with the Spanish test plant, the two outcomes matched very well. Later on, the model was used to predict the performance of large-scale SCPPS, the predicted results showed that the chimney height, the turbine pressure drop factor, the diameter of the collector and its optical properties were important parameters of SCPPS.

Serag-Eldin [168] established a calculation model to predict the fluid flow in the SCPPS. The model was made up of the mass, momentum and energy conservation equations, and two turbulence equations. Furthermore, for the first time, he proposed that we should use an actuator dish model to describe the effect of turbine. The model can be used to estimate the influence of the system size on the system output power. Pastohr [169] simulated the Spanish solar chimney test power plant, established heat and flow mathematical model in the heat collector, the heat storage layer, the turbine, and the chimney. The author made two important simplifications: (1) The heat storage layer was regarded as solid; (2) The turbine was considered as a reverse fan. Turbine pressure drop was calculated based on the Betz theory. Comparing the steady-state numerical calculation with the SIMPLE calculation results, detailed improving measures for the mathematical model were put forward. Moreover, Pretorius and Kröger [170] and Bilgen and Rheault [171] established the corresponding mathematical model for SCPPS respectively. They considered the influences of the following factors: solar radiation, earth latitude, and convective heat transfer coefficient, on the SCPP performance. Coetzee [172] established a transient heat transfer mathematical model for the heat collector, chimney, and turbine, considering the influence of solar radiation on the system output power. Furthermore, he designed a new nozzle shape chimney structure and pyramid-shape collector structure and made the design optimization.

The group of von Backström established a compressible air flow heat transfer mathematical model inside the chimney of the SCPPS [173,174]. Research suggested that, compared with inlet air density, chimney outlet air density decreased significantly. A one-dimensional compressible fluid dynamics model was established to describe the influence of factors, such as the chimney height, wall friction, the internal structure, local resistance loss, sudden shrinkage and enlargement in flow channel, on the thermodynamic parameters of the system. They also analyzed the influence of the Mach number, the fluid density and flow velocity on the pressure change inside the chimney. Furthermore, they predicted various energy loss in the large-scale SCPPS and pressure change at the inlet and outlet of the chimney. Finally, they provided a detailed data analysis report for the fluid flow characteristics inside the chimney.

1.3.2.4 The operation principle of the turbine and the design optimization techniques

As early as 1980, Haaf et al. [114] and Schlaich [113] had carried out the preliminary research on the flow mechanism of the axial turbine. In 1985, Kustrin and Tuma [175] pointed out that the installed wind turbine at the bottom of the chimney shall be single stage and based on the pressure type. Only a part of the air flow kinetic energy could be converted to electrical power. Later on, he described the fundamental principle of the turbine and made a calculation to optimize the test plant in Spain on the basis of the wind turbine analysis.

The research group coming from Professor Backström in South Africa [159,176−181] carried out detailed analysis of the characteristic parameters of wind turbine, flow channel design and optimization, the output power, and so on, as well as vast experimental research. Their main work was as follows: (1) They developed the single rotor and introduced the chimney inlet guide vane that induced fluid flow prerotation before entering the turbine. This design helped to reduce the kinetic energy loss at the entrance of the rotor. They gave the performance and efficiency of the two groups of operating experiments with the similar model of turbines. The measurement showed that the design total-to-total efficiency could be increased up to 85−90%, and the total-to-static efficiency could be increased up to 77−80%. (2) They established the output power mathematical model based on fluid flow of the turbine, load factor and degree of reaction. The optimum reaction, the output power, and turbine size for the largest turbine efficiency were obtained by numerical simulations. The measurement results for a 720 mm turbine showed the correctness of the theoretical analysis of the model. For a given SCPPS, experiments demonstrated that the largest turbine total-to-total efficiency can be increased up to 90%. (3) They made a fluid flow numerical simulation in the turbine area of SCPPS, including the inlet guide vane, the turbine set, outlet of the collector and the inlet of chimney regions. Then they analyzed the influence of wall resistance coefficient, the collector height, the chimney diameter, the turbine diameter, and turbine leaf shape on the pressure drop between the inlet and outlet of the chimney. The experimental results, including the internal flow angle, velocity component, and wall static pressure, had good consistency with the prediction made by commercial CFD code.

1.3.3 ECONOMIC AND ECOLOGICAL THEORY AND FEASIBILITY STUDIES

Since SCPPS gets involved in many important research areas, such as energy, environment, electricity, and construction engineering, research on its economics and feasibility has never stopped. One of the main reasons was the low overall efficiency of the system. Some associated researchers in the international community had questioned the feasibility of the system. But now, the international community has reached a consensus on the economic feasibility of the system, that is, in terms of total investment costs, operating costs, environmental friendliness, etc., the SCPPS is economical and feasible.

Since the 1980s, Schlaich et al. [109,113−116,125,150] have tracked reports on the economy and feasibility of SCPPS. They analyzed the impact of the world current energy situation, environment, and population on the economy. They proposed the construction of SCPPS was an important measure to solve energy problem in the arid and semi-arid areas in undeveloped countries and regions. Mullet [152], Lodhi [155,157,162], Stinnes [182], Dai et al. [183], Onyango et al. [184] analyzed and studied the economics and feasibility of the system. They thought that the construction of SCPPS in underdeveloped areas and deserts was efficient and technically feasible. Beerbaum and Weinrebe [185] made a profound comparison and analysis among the economy of the coal power generation, solar trough thermal power generation, tower thermal power generation, dish thermal power generation, and solar chimney power generation technologies. They gave a detailed analysis of India's energy situation and solar distribution, and proposed that developing the SCPPS was the most economical and feasible solution for the energy shortage in India. The technology could be vigorously promoted in the Indian desert and underdeveloped regions.

Undoubtedly, SCPPS will bring excellent ecological and environmental effects. Lodhi [157] and Bernardes [186] respectively conducted an in-depth analysis of the environmental impact and life cycle assessment for SCPPS. Compared with the coal power plant, SCPPS can reduce the amount of SO_x, NO_x, and CO_2 emissions, which is an important measure to improve the urban environment and alleviate environmental pollution around the world.

1.3.4 POTENTIAL APPLICATION OF SCPPS

The development of SCPPS was nagged by the low efficiency, complicated design, high initial investment, and huge risks. Based on the actual situation, researchers put forward some new ideas to improve the system.

Michaud proposed the use of hurricane updraft to drive a turbine to generate electricity [153,187]. Papageorgiou's research group [188−193] proposed a floating solar chimney system, its main purpose was to solve the problems in the construction of the chimney (Fig. 1.21). The research team carried out a series of studies on the floating solar chimney system. They analyzed the thermodynamic cycle, the heat transfer and fluid flow characteristics, output power, the design and optimization of the turbine, and the economic and technical feasibility of the system. Serag-Eldin [194−196], Zhou et al. [197], Cao et al. [198−200], Koonsrisuk [201,202], Kalash et al. [146,147,203], and Li et al. [204,205] presented mathematical models to describe the operation of sloping SCPPS. Many researchers considered using SCPPS for seawater desalination [68,206−212]. Since the 1990s, the design of SCPPS has entered a substantive development stage. Australia, South Africa, India, Egypt, and some other countries have shown

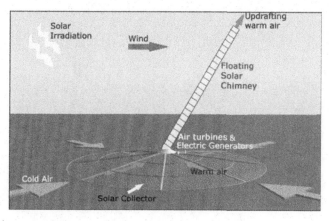

FIGURE 1.21

Floating solar chimney power plant in operation [193].

great interest in the construction of a particular scale SCPPS to solve the power shortage problem. They believe that human large-scale exploitation of productive solar energy resource in the desert area using cheap and available construction materials was going to be achieved gradually. The plan and design of 100 MW and 200 MW power plants will be realized in the coming years. The construction of a 30 MW solar chimney power plant in Thar Desert of Rajasthan [213], India was proved to be feasible and began to be implemented. But because of the nuclear arms race between India and Pakistan, the plans came to nothing. Allegedly, the power plant location was considered to be difficult to defend against deliberate destruction, leaving the project to be canceled.

Under the financial support of South Africa's Northern Cape province, Stinnes and his research team have been conducting feasibility studies of SCPPS since 1995 [182]. They planned to build a practical scale SCPPS in Sishen, a remote desert town in South Africa. It is an ambitious program with a solar chimney height of up to 1500 m, a height three times that of the CN Tower in Canada. The diameter of the collector is 4 km. The project was expected to cost 250 million pounds, and power generation capacity was 200 MW. Stinnes compared this cheap, clean power plant with a variety of other power plants. If the power plant successfully operated for 80 years, it was equivalent to two and a half coal-fired power plants or four combined-cycle natural gas power plant with the same power capacity. Even in the worst case, the electricity price of SCPPS was equal to that of the conventional power plants. In the best case, however, the electricity price of SCPPS would be one-third of that of the conventional power plants.

The world's first large-scale SCPPS with a height of 1000 m is planned to start construction in Mildura, Australia [214], with the support and attention of the federal government (Fig. 1.22). The chief architect of the project was Professor Jörg Schlaich and the planned investment was 700 million Australian dollars (US \$395 million). The diameter of the plant collectors reaches 7 km,

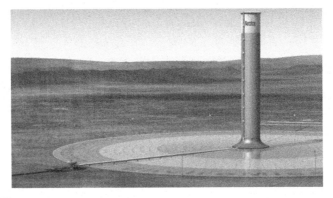

FIGURE 1.22

EnviroMission's solar chimney power plant in Mildura, Australia [214].

32 turbine generator units with rated output power being 6.25 MW would be installed under the chimney. The total annual power generation capacity would be 200 MW, which could meet the living electricity demand for 200,000 residents.

1.4 RESEARCH CONTENTS OF THIS BOOK

Based on the background of the SCPPS, this book will focus on various aspects, such as the further research of the thermodynamic theory, HAG effect, impacts of structural parameters on fluid flow and heat transfer characteristics, energy storage performance, heat and mass transfer characteristics of the SCPP coupled with turbine, SCPPS under ambient crosswind, and experimental results of small-scale demonstration models. Specifically, the main contents for the following sections of this book are as follows.

1. Firstly, the thermodynamic process in all the essential components of SCPPS is analyzed. By analyzing thermodynamic parameters in each process, the thermodynamic cycle in SCPPS is reestablished. Then the maximum thermal to electric energy conversion efficiency of the system is found. Later, the standard Brayton cycle efficiency, system maximum energy conversion efficiency, and output power of the different for different scale SCPPSs is obtained and compared.

2. A comprehensive mathematical model is advanced to evaluate the HAG effect of an SCPPS, in which the effects of various parameters on the relative static pressure, driving force, power output, and efficiency are further investigated as the existing models are insufficient to accurately describe the driving mechanism of SCPPS. Using the SCPPS prototype in Manzanares, Spain as a practical example, numerical studies are performed to explore the geometric modifications on the system performance, which show reasonable agreement with the analytical model.

3. Numerical simulations on air flow, heat transfer, and power output characteristics of an SCPPS with energy storage layer and turbine similar to the Spanish prototype are carried out, and a mathematical model of flow and heat transfer for the solar chimney power plant system is established. The influences of solar radiation and pressure drop across the turbine on the flow and heat transfer, output power, and energy loss of the solar chimney power plant system are analyzed. In addition, the large mass flow rate of air flowing through the chimney outlet becomes the main cause of energy loss in the system, and the collector canopy also results in large energy loss. Further, the influences of the storage layer material, solar radiation, and environment wind on the system performance are analyzed; the impacts of the heat storage medium characteristics on the power generation peak-valley difference are studied.

4. Numerical simulations are carried out on the SCPPS systems coupled with the turbine. The whole system is divided into three regions: the collector, the chimney, and the turbine, and the mathematical models of heat transfer, and flow have been set up for these regions. Using the Spanish prototype as a practical example, numerical simulation results for the prototype with a 3-blade turbine show that the maximum power output of the system is a little higher than 50 kW. Furthermore, the effect of the turbine rotational speed on the chimney outlet parameters has been analyzed which shows the validity of the numerical method. Thereafter, design and simulation of a MW-graded solar chimney power plant system with a 5-blade turbine are presented, and the numerical simulation results show that the power output and turbine efficiency are 10 MW and 50%, respectively, which presents a reference to the design of large-scale solar chimney power plant systems.

5. Numerical simulations were carried out to analyze the energy storage performance of the solar chimney power plant systems with an energy storage layer. Mathematical models were developed to describe the flow and heat transfer mechanisms of the collector, chimney, and energy storage layer, and the response of different energy storage materials to the solar radiation and the effects of these materials on the power output with different solar radiation were analyzed.

6. Ambient crosswind has a complicated influence on the performance of SCPPS both via the collector inlet and via the chimney outlet. To investigate the impact of strong ambient crosswinds on the system output power through the collector inlet and chimney outlet, numerical analysis on the performances of a SCPPS identical to the prototype in Manzanares, Spain, which is exposed to the external crosswind with different velocities, is carried out. A geometrical model including the SCPPS and its outside ambience is built, and the mathematical models to describe the fluid flow, heat transfer, and output power of the whole system are further developed. The pressure, temperature, and velocity distribution of the air in the ambience and SCPPS together with the output power of the SCPPS are analyzed. To overcome the negative impact of strong ambient crosswind on the system performance through the collector inlet, a 2-m high blockage was circularly set a few meters away from the collector inlet.

7. A mini-scale solar chimney prototype has been set up, and the temperature distribution of the system with time and space, together with the velocity variation inside the chimney with time, has been measured. The experimental results indicate that the temperature distributions inside the collector and the effects of seasons on the heat transfer and flow characteristics of the system show great agreement with the analysis, while the temperature decreases significantly inside the chimney as the chimney is very thin which causes very high heat loss.

8. A brief introduction on the future research points of the SCPPS has been presented from the basic theories to the commercial applications.

REFERENCES

[1] US EIA. Top world oil producers in 2013, <http://www.eia.gov/countries/>; 2015.

[2] US Environmental Protection Agency. Overview of greenhouse gases, <http://www.epa.gov/climatechange/ghgemissions/gases/co2.html>; 2015.

[3] Hance J. Record year for CO_2 emissions, even with economic slowdown, <http://news.mongabay.com/2009/1117-hance_carbonemissions.html>; 2009.

[4] Moog HC, Bok F, Marquardt CM, Brendler V. Disposal of nuclear waste in host rock formations featuring high-saline solutions — implementation of a thermodynamic reference database (THEREDA). Appl Geochem 2015;55:72—84.

[5] Kalmykov SN, Vlasova IE, Romanchuk AY, Zakharova EV, Volkova AG, Presnyakov IA. Partitioning and speciation of Pu in the sedimentary rocks aquifer from the deep liquid nuclear waste disposal. Radiochim Acta 2015;103:175—85.

[6] Zheng LE, Rutqvist J, Liu HH, Birkholzer JT, Sonnenthal E. Model evaluation of geochemically induced swelling/shrinkage in argillaceous formations for nuclear waste disposal. Appl Clay Sci 2014;97—98:24—32.

[7] Yli-Kauhaluoma S, Hanninen H. Tale taming radioactive fears: linking nuclear waste disposal to the "continuum of the good". Public Underst Sci 2014;23:316—30.

[8] Rechard RP, Wilson ML, Sevougian SD. Progression of performance assessment modeling for the Yucca Mountain disposal system for spent nuclear fuel and high-level radioactive waste. Reliab Eng Syst Safe 2014;122:96—123.

[9] Rechard RP, Freeze GA, Perry FV. Hazards and scenarios examined for the Yucca Mountain disposal system for spent nuclear fuel and high-level radioactive waste. Reliab Eng Syst Safe 2014;122:74—95.

[10] Moyce EBA, Rochelle C, Morris K, Milodowski AE, Chen XH, Thornton S, et al. Rock alteration in alkaline cement waters over 15 years and its relevance to the geological disposal of nuclear waste. Appl Geochem 2014;50:91—105.

[11] Kari OP, Puttonen J. Simulation of concrete deterioration in Finnish rock cavern conditions for final disposal of nuclear waste. Ann Nucl Energy 2014;72:20—30.

[12] Schilling DR. In depth: global energy demand 2035 and integrated global systems model (IGSM), <http://www.industrytap.com/will-earth-keep-up-with-increasing-energy-demands/760>; 2013.

[13] He W, Zhou JZJ, Hou X, Chen C, Ji J. Theoretical and experimental investigation on a thermoelectric cooling and heating system driven by solar. Appl Energ 2013;107:89—97.

[14] Gasik M, Bilotsky Y. Optimisation of functionally gradated material thermoelectric cooler for the solar space power system. Appl Therm Eng 2014;66:528—33.

[15] Chen WH, Wang CC, Hung CI, Yang CC, Juang RC. Modeling and simulation for the design of thermal-concentrated solar thermoelectric generator. Energy 2014;64:287—97.

[16] Liu ZB, Zhang L, Gong GC, Han TH. Experimental evaluation of an active solar thermoelectric radiant wall system. Energ Convers Manage 2015;94:253—60.

[17] He W, Su YH, Wang YQ, Riffat SB, Ji J. A study on incorporation of thermoelectric modules with evacuated-tube heat-pipe solar collectors. Renew Energ 2012;37:142—9.

[18] Kalogirou SA. Solar thermoelectric power generation in Cyprus: selection of the best system. Renew Energ 2013;49:278—81.

[19] Date A, Date A, Dixon C, Akbarzadeh A. Theoretical and experimental study on heat pipe cooled thermoelectric generators with water heating using concentrated solar thermal energy. Sol Energy 2014;105:656—68.

[20] Zhang CB, Najafi K, Bernal LP, Washabaugh PD. Micro combustion-thermionic power generation: feasibility, design and initial results. In: Boston transducers '03: digest of technical papers, 1 and 2; 2003, p. 40-44.

[21] Zeng T. Multi-layer thermionic-tunneling structures for power generation. Adv Electron Packaging 2005;Pts A—C:2181—3.

[22] Yarygin VI, Ionkin VI, Kuptsov GA, Ovcharenko MK, Ruzhnikov VA, Pyshko AP, et al. New-generation space thermionic nuclear power systems with out-of-core electricity generating systems. Atom Energy 2000;89:528—40.

[23] Vashaee D, Shakouri A. Thermionic power generation at high temperatures using SiGe/Si superlattices. J Appl Phys 2007;101.

[24] Zeng TF. Thermionic-tunneling multilayer nanostructures for power generation. Appl Phys Lett 2006;88.

[25] Liu R, Zhang TL, Yang L, Zhou ZN. Effect of particle size on thermal decomposition of alkali metal picrates. Thermochim Acta 2014;583:78—85.

[26] Kolezynski A, Malecki A. Theoretical studies of electronic structure and structural properties of anhydrous alkali metal oxalates part II. Electronic structure and bonding properties versus thermal decomposition pathway. J Therm Anal Calorim 2014;115:841—52.

[27] Kaur K, Kaur J, Arora B, Sahoo BK. Emending thermal dispersion interactions of Li, Na, K, and Rb alkali-metal atoms with graphene in the Dirac model. Phys Rev B 2014;90.

[28] Ishii Y, Sato K, Salanne M, Madden PA, Ohtori N. Thermal conductivity of molten alkali metal fluorides (LiF, NaF, KF) and their mixtures. J Phys Chem B 2014;118:3385—91.

[29] Fijalkowski KJ, Jaron T, Leszczynski PJ, Magos-Palasyuk E, Palasyuk T, Cyranski MK, et al. M(BH3NH2BH2NH2BH3) — the missing link in the mechanism of the thermal decomposition of light alkali metal amidoboranes. Phys Chem Chem Phys 2014;16:23340—6.

[30] Kumar MVS, Rajesh D, Balakrishna A, Ratnakaram YC. Thermal and optical properties of Nd3+ doped lead zinc borate glasses-influence of alkali metal ions. Physica B 2013;415:67—71.

[31] Przybylski W, Grybos R, Majda D, Szklarzewicz JT. Effect of alkali metal ion and hydrogen bonds on thermal stability of M[VO(O-2)(2)bpy]center dot nH(2)O (M = Li + -Rb +) and Cs[VO(O-2)(2)bPY]center dot H2O2 complexes. Thermochim Acta 2011;514:32—6.

[32] Lee J, Lee JS, Kang M. Synthesis of group IA alkali metal-aluminosilicates, and their hydrogen production abilities on methanol thermal decomposition. Energy 2011;36:3293—301.

[33] Sakai T, Matsumoto M, Murakami T, Okuno Y. Numerical simulation of power generation characteristics of a disk MHD Generator with high-temperature inert gas plasma. Electr Eng Jpn 2012;179:23—30.

[34] Lee KB, Rhi SH, Lee KW, Lee WH, Jang CC, Lee WG, et al. Thermal and flow modeling of alkali-metal thermoelectric power generation (AMTEC). In: Mater proc technol II, Pts 1—4, 538—541; 2012. p. 419—22.

[35] Ohno J, Liberati A, Murakami T, Okuno Y. Numerical study of plasma-fluid behavior and generation characteristics of the closed-loop MHD electrical power generator. Electr Eng Jpn 2011;174:37—44.

[36] Li YW, Li YH, Lu HY, Zhu T, Zhang BL, Chen F, et al. Preliminary experimental investigation on MHD power generation using seeded supersonic argon flow as working fluid. Chinese J Aeronaut 2011;24:701−8.

[37] Wang XL, Ye D, Gu F. Investigation of the characteristics of MHD power generation by a corona jet across a magnetic field. IEEE T Plasma Sci 2008;36:299−304.

[38] Murakami T, Okuno Y, Yamasaki H. Non-equilibrium plasma MHD electrical power generation at Tokyo tech. Complex Syst 2008;982:618−25.

[39] Pak PS, Suzuki Y, Kosugi T. A CO_2-capturing hybrid power-generation system with highly efficient use of solar thermal energy. Energy 1997;22:295−9.

[40] Sukhatme SP. Solar thermal power generation. P Indian as-Chem Sci 1997;109:521−31.

[41] Gupta MK, Kaushik SC. Exergy analysis and investigation for various feed water heaters of direct steam generation solar-thermal power plant. Renew Energ 2010;35:1228−35.

[42] Hu E, Yang YP, Nishimura A, Yilmaz FA. Kouzani, solar thermal aided power generation. Appl Energ 2010;87:2881−5.

[43] Shabgard H, Bergman TL, Faghri A. Exergy analysis of latent heat thermal energy storage for solar power generation accounting for constraints imposed by long-term operation and the solar day. Energy 2013;60:474−84.

[44] Jones BW, Powell R. Evaluation of distributed building thermal energy storage in conjunction with wind and solar electric power generation. Renew Energ 2015;74:699−707.

[45] Sharma C, Sharma AK, Mullick SC, Kandpal TC. Assessment of solar thermal power generation potential in India. Renew Sust Energ Rev 2015;42:902−12.

[46] Zhang HL, Wang ZF, Guo MH, Liang WF. Cosine efficiency distribution of heliostats field of solar thermal power tower plants. Asia-Pac Power Energ 2009:623−6.

[47] Wang ZF, Yao ZH, Dong J, Jin HG, Han W, Lu ZW, et al. The design of a 1MW solar thermal tower plant in Beijing, China. In: Proceedings of ISES solar world congress 2007: solar energy and human settlement, vols I−V; 2007. p. 1729−32.

[48] Reddy VS, Kaushik SC, Tyagi SK. Exergetic analysis and economic evaluation of central tower receiver solar thermal power plant. Int J Energ Res 2014;38:1288−303.

[49] Xu ES, Yu QA, Wang ZF, Yang CY. Modeling and simulation of 1 MW DAHAN solar thermal power tower plant. Renew Energ 2011;36:848−57.

[50] Flueckiger SM, Iverson BD, Garimella SV, Pacheco JE. System-level simulation of a solar power tower plant with thermocline thermal energy storage. Appl Energ 2014;113:86−96.

[51] Price H, Lupfert E, Kearney D, Zarza E, Cohen G, Gee R, et al. Advances in parabolic trough solar power technology. J Sol Energ-T Asme 2002;124:109−25.

[52] Bradshaw RW, Siegel NP. Molten nitrate salt development for thermal energy storage in parabolic trough solar power systems, Es2008. In: Proceedings of the 2nd international conference on energy sustainability, vol. 2; 2009. p. 631−7.

[53] Hou HJ, Yang YP, Hu E, Song JF, Dong CQ, Mao J. Evaluation of solar aided biomass power generation systems with parabolic trough field. Sci China Technol Sci 2011;54:1455−61.

[54] Channon SW, Eames PC. The cost of balancing a parabolic trough concentrated solar power plant in the Spanish electricity spot markets. Sol Energy 2014;110:83−95.

[55] Poghosyan V, Hassan MI. Techno-economic assessment of substituting natural gas based heater with thermal energy storage system in parabolic trough concentrated solar power plant. Renew Energ 2015;75:152−64.

[56] Sait HH, Martinez-Val JM, Abbas R, Munoz-Anton J. Fresnel-based modular solar fields for performance/cost optimization in solar thermal power plants: a comparison with parabolic trough collectors. Appl Energ 2015;141:175–89.

[57] Trad A, Ali MAA. Determination of the optimum design through different funding scenarios for future parabolic trough solar power plant in Algeria. Energ Convers Manage 2015;91:267–79.

[58] Kaneff S. Viable distributed dish central plant solar power: status, new developments, potential. J Phys Iv 1999;9:195–200.

[59] Cavallaro F, Ciraolo L. A life cycle assessment (LCA) of a paraboloidal-dish solar thermal power generation system, 2006 First international symposium on environment identities and Mediterranean area, vols. 1 and 2; 2006. p. 127–32.

[60] Abbas M, Boumeddane B, Said N, Chikouche A. Dish stirling technology: a 100 MW solar power plant using hydrogen for Algeria. Int J Hydrogen Energ 2011;36:4305–14.

[61] Eswaramoorthy M, Shanmugam S. Solar parabolic dish thermoelectric generator: a technical study. Energ Source Part A 2013;35:487–94.

[62] Muthu G, Shanmugam S, Veerappan AR. Solar parabolic dish thermoelectric generator with acrylic cover. In: 4 international conference on advances in energy research (ICAER 2013); 2014(54). p. 2–10.

[63] Shanmugam S, Veerappan A, Eswaramoorthy M. An experimental evaluation of energy and exergy efficiency of a solar parabolic dish thermoelectric power generator. Energ Source Part A 2014;36:1865–70.

[64] Zhang S, Wu ZH, Zhao RD, Yu GY, Dai W, Luo EC. Study on a basic unit of a double-acting thermoacoustic heat engine used for dish solar power. Energ Convers Manage 2014;85:718–26.

[65] Hu YS, Yan Q, Yang YP. Economic analysis of solar trough, tower and dish power plants. Energy Power Technol, Pts 1 and 2 2013;805–806:12–16.

[66] Alonzo CA, Garcia W, Saloma C. Crosstalk between two-photon and two-color (two-photon) excitation in optical beam induced current generation with two confocal excitation beams. Opt Commun 2007;270:139–44.

[67] Chen WY, Mattern DL, Okinedo E, Senter JC, Mattei AA, Redwine CW. Photochemical and acoustic interactions of biochar with CO_2 and H_2O: applications in power generation and CO_2 capture. Aiche J 2014;60:1054–65.

[68] Strack G, Luckarift HR, Sizemore SR, Nichols RK, Farrington KE, Wu PK, et al. Power generation from a hybrid biological fuel cell in seawater. Bioresour Technol 2013;128:222–8.

[69] Bin H. Research on biological greenhouse photovoltaic power generation access system design. In: 2013 international conference on industrial engineering and management science (ICIEMS 2013); 2013. p. 957–64.

[70] Hosenuzzaman M, Rahim NA, Selvaraj J, Hasanuzzaman M, Malek ABMA, Nahar A. Global prospects, progress, policies, and environmental impact of solar photovoltaic power generation. Renew Sust Energ Rev 2015;41:284–97.

[71] Gunderson I, Goyette S, Gago-Silva A, Quiquerez L, Lehmann A. Climate and land-use change impacts on potential solar photovoltaic power generation in the Black Sea region. Environ Sci Policy 2015;46:70–81.

[72] Wang Y, Zhou S, Huo H. Cost and CO_2 reductions of solar photovoltaic power generation in China: perspectives for 2020. Renew Sust Energ Rev 2014;39:370–80.

[73] Singh GK. Solar power generation by PV (photovoltaic) technology: a review. Energy 2013;53:1−13.

[74] Pacas JM, Molina MG, dos Santos EC. Design of a robust and efficient power electronic interface for the grid integration of solar photovoltaic generation systems. Int J Hydrogen Energ 2012;37:10076−82.

[75] Pearce JM. Expanding photovoltaic penetration with residential distributed generation from hybrid solar photovoltaic and combined heat and power systems. Energy 2009;34:1947−54.

[76] Ahmed NA, Miyatake M, Al-Othman AK. Power fluctuations suppression of stand-alone hybrid generation combining solar photovoltaic/wind turbine and fuel cell systems. Energ Convers Manage 2008;49:2711−19.

[77] Contreras A, Guirado R, Veziroglu TN. Design and simulation of the power control system of a plant for the generation of hydrogen via electrolysis, using photovoltaic solar energy. Int J Hydrogen Energ 2007;32:4635−40.

[78] Quaschning V. Technical and economical system comparison of photovoltaic and concentrating solar thermal power systems depending on annual global irradiation. Sol Energy 2004;77:171−8.

[79] Trieb F, Kronshage S, Knies G. Concentrating on solar power in a trans-Mediterranean renewable energy co-operation. In: Proceedings of the 4th international conference on solar power from space − SPS '04, together with the 5th international conference on wireless power transmission − WPT 5; 2004(567), p. 99−108.

[80] Van Voorthuysen E.H.D.M. The promising perspective of concentrating solar power (CSP). In: 2005 international conference on future power systems (FPS); 2005. p. 47−53.

[81] Pitz-Paal R, Dersch J, Milow B, Tellez F, Ferriere A, Langnickel U, et al. Development steps for concentrating solar power technologies with maximum impact on cost reduction − results of the European ECOSTAR study. Solar Eng 2005;2006:773−9.

[82] Hang Q, Jun Z. Feasibility and potential of concentrating solar power in China. In: Proceedings of ISES solar world congress 2007: solar energy and human settlement, vols. I−V; 2007. p. 1724−8.

[83] Hennecke K, Schwarzbozl P, Hoffschmidt B, Gottsche J, Koll G, Beuter M, et al. The solar power tower julich − a solar thermal power plant for test and demonstration of air receiver technology. In: Proceedings of ISES solar world congress 2007: solar energy and human settlement, vols. I−V; 2007. p. 1749−53.

[84] Vant-Hull LL. Concentrating solar thermal power (CSP). In: Proceedings of ISES solar world congress 2007: solar energy and human settlement, vols. I−V; 2007. p. 68−74.

[85] Bode CC, Sheer TJ. A techno-economic feasibility study on the use of distributed concentrating solar power generation in Johannesburg. J Energy South Afr 2010;21:2−11.

[86] Wikipedia, Solar power plants in the Mojave Desert, <http://en.wikipedia.org/wiki/Solar_power_plants_in_the_Mojave_Desert>; 2015.

[87] Tian YL, Shao YL, Lu P, Cheng JS, Liu WC. Effect of SiO2/B2O3 ratio on the property of borosilicate glass applied in parabolic trough solar power plant. J Wuhan Univ Technol 2015;30:51−5.

[88] Peng S, Hong H, Jin HG, Zhang ZN. A new rotatable-axis tracking solar parabolic-trough collector for solar-hybrid coal-fired power plants. Sol Energy 2013;98:492−502.

[89] Bakos GC, Tsechelidou C. Solar aided power generation of a 300 MW lignite fired power plant combined with line-focus parabolic trough collectors field. Renew Energ 2013;60:540−7.

[90] Bakos GC, Parsa D. Technoeconomic assessment of an integrated solar combined cycle power plant in Greece using line-focus parabolic trough collectors. Renew Energ 2013;60:598−603.

[91] Reddy VS, Kaushik SC, Tyagi SK. Exergetic analysis and performance evaluation of parabolic trough concentrating solar thermal power plant (PTCSTPP). Energy 2012;39:258−73.

[92] Palenzuela P, Zaragoz G, Alarcon-Padilla DC, Guillen E, Ibarra M, Blanco J. Assessment of different configurations for combined parabolic-trough (PT) solar power and desalination plants in arid regions. Energy 2011;36:4950−8.

[93] Pitz-Paal R, Dersch J, Milow B, Tellez F, Ferriere A, Langnickel U, et al. Development steps for parabolic trough solar power technologies with maximum impact on cost reduction. J Sol Energ-T Asme 2007;129:371−7.

[94] Michels H, Pitz-Paal R. Cascaded latent heat storage for parabolic trough solar power plants. Sol Energy 2007;81:829−37.

[95] Herrmann U, Kelly B, Price H. Two-tank molten salt storage for parabolic trough solar power plants. Energy 2004;29:883−93.

[96] Wikipedia, World energy consumption, <http://en.wikipedia.org/wiki/World_energy_consumption>; 2015.

[97] Li Y, Choi SS, Yang C, Wei F. Design of variable-speed dish-stirling solar-thermal power plant for maximum energy harness. IEEE T Energy Conver 2015;30:394−403.

[98] Reddy KS, Veershetty G. Viability analysis of solar parabolic dish stand-alone power plant for Indian conditions. Appl Energ 2013;102:908−22.

[99] Bakos GC, Antoniades C. Techno-economic appraisal of a dish/stirling solar power plant in Greece based on an innovative solar concentrator formed by elastic film. Renew Energ 2013;60:446−53.

[100] Li ZG, Tang DW, Du JL, Li T. Study on the radiation flux and temperature distributions of the concentrator-receiver system in a solar dish/stirling power facility. Appl Therm Eng 2011;31:1780−9.

[101] Wu SY, Xiao L, Cao YD, Li YR. A parabolic dish/AMTEC solar thermal power system and its performance evaluation. Appl Energ 2010;87:452−62.

[102] Wu SY, Xiao L, Cao YD, Li YR. Convection heat loss from cavity receiver in parabolic dish solar thermal power system: a review. Sol Energy 2010;84:1342−55.

[103] Li X, Wang ZF, Yu J, Liu XB, Li J, Song XO. The power performance experiment of Dish-Stirling solar thermal power system. In: Proceedings of ISES solar world congress 2007: solar energy and human settlement, vols. I−V; 2007: p. 1858−62.

[104] Wikipedia, Concentrated solar power, <http://en.wikipedia.org/wiki/Concentrated_solar_power>; 2015.

[105] Silveira JL, Tuna CE, Lamas WD. The need of subsidy for the implementation of photovoltaic solar energy as supporting of decentralized electrical power generation in Brazil. Renew Sust Energ Rev 2013;20:133−41.

[106] Gadonneix P, Sambo A, Nadeau MJ, Statham BA, Kim YD, Birnbaum L, Lleras JAV, Cho H-E, Graham Ward C. World energy resources 2013 survey,

World Energy Council, <http://www.worldenergy.org/wp-content/uploads/2013/09/Complete_WER_2013_Survey.pdf>; 2013.

[107] Wikipedia, Solar updraft tower, <http://en.wikipedia.org/wiki/Solar_updraft_tower>; 2015.

[108] Wikipedia, Isidoro Cabanyes, <http://es.wikipedia.org/wiki/Isidoro_Cabanyes>; 2014.

[109] J S. G M. W H. Aufwindkraftwerke-Die demonstrationsanlage in MANZANARES/SPANIEN. (Upwind power plants-the demonstration plant in Manzanares, Spain). In: Proceedings of the national conference on power transmission; 1980. p. 97−112.

[110] Robert R. Spanish solar chimney nears completion. MPS Rev 1981:21−3.

[111] Robert R. Solar prototype development in Spain show great promise. MPS Rev 1982:21−3.

[112] Robert R. Hot air starts to rise through Spain's solar chimney. Elect Rev 1982;210:26−7.

[113] Schlaich J. Solar chimneys. Periodica 1983:45.

[114] Haaf W, Friedrich K, Mayer G, Schlaich J. Solar chimneys. Int J Solar Energy 1983;2:3−20.

[115] Haaf W, Friedrich K, Mayer G, Schlaich J. Solar chimneys. Int J Solar Energy 1984;2:141−61.

[116] Schlaich J. The solar chimney, edition axel menges. Stuttgart; 1995.

[117] Cocco D, Serra F. Performance comparison of two-tank direct and thermocline thermal energy storage systems for 1 MWe class concentrating solar power plants. Energy 2015;81:526−36.

[118] Nithyanandam K, Pitchumani R, Mathur A. Analysis of a latent thermocline storage system with encapsulated phase change materials for concentrating solar power. Appl Energ 2014;113:1446−60.

[119] Nithyanandam K, Pitchumani R. Cost and performance analysis of concentrating solar power systems with integrated latent thermal energy storage. Energy 2014;64:793−810.

[120] Liu M, Belusko M, Tay NHS, Bruno F. Impact of the heat transfer fluid in a flat plate phase change thermal storage unit for concentrated solar tower plants. Sol Energy 2014;101:220−31.

[121] Flueckiger SM, Garimella SV. Latent heat augmentation of thermocline energy storage for concentrating solar power − a system-level assessment. Appl Energ 2014;116:278−87.

[122] Al-Soud MS, Hrayshat ES. A 50 MW concentrating solar power plant for Jordan. J Clean Prod 2009;17:625−35.

[123] de_Richter R. Réacteurs Météorologiques, <http://www.tour-solaire.fr/publications-scientifiques.php>; 2015.

[124] Günther H. In hundert Jahren − Die künftige Energieversorgung der Welt (In hundred years − future energy supply of the world). Stuttgart: Kosmos, Franckh'sche Verlagshandlung; 1931.

[125] Lautenschlager H, Haff HJS. New results from the solar-chimney prototype and conclusions for larger plants. In: European wind energy conference Hamburg; 1984. p. 231−5.

[126] Krisst R. Energy transfer system. Altern Sources Energy 1983;63:8−11.

[127] Kulunk H. A prototype solar convection chimney operated under Izmit condition. In: Veziroglu TN, editor. Prod. 7th MICAS; 1985. p. 162.

[128] Padki MM, Sherif SA. Fluid dynamics of solar chimneys. In: Morrow TB, Marshall LR, Simpson RL, editors. Forum on industrial applications of fluid mechanics, FED-vol. 70. New York: ASME; 1988. p. 43−6.

[129] Padki MM, Sherif SA. Solar chimney for medium-to-large scale power generation. In: Proceedings of the manila international symposium on the development and management of energy resources; 1989. p. 432.

[130] Padki MM, Sherif SA. A mathematical model for solar chimneys. In: Proceedings of 1992 international renewable energy conference. Amman, Jordan; 1992. p. 289−94.

[131] Pasumarthi N, Sherif SA. Experimental and theoretical performance of a demonstration solar chimney model − part II: experimental and theoretical results and economic analysis. Int J Energ Res 1998;22:443−61.

[132] Pasumarthi N, Sherif SA. Experimental and theoretical performance of a demonstration solar chimney model − part I: mathematical model development. Int J Energ Res 1998;22:277−88.

[133] Padki MM, Sherif SA. On a simple analytical model for solar chimneys. Int J Energ Res 1999;23:345−9.

[134] Akbarzadeh A, Johnson P, Singh R. Examining potential benefits of combining a chimney with a salinity gradient solar pond for production of power in salt affected areas. Sol Energy 2009;83:1345−59.

[135] Zhao YC, Akbarzadeh A, Andrews J. Combined water desalination and power generation using a salinity gradient solar pond as a renewable energy source. In: Proceedings of ISES solar world congress 2007: solar energy and human settlement, vols. I−V; 2007. p. 2184−8.

[136] Zhou XP, Yang JK, Xiao B, Hou GX. Experimental study of temperature field in a solar chimney power setup. Appl Therm Eng 2007;27:2044−50.

[137] Zhou XP, Yang JK, Xiao B, Hou GX. Simulation of a pilot solar chimney thermal power generating equipment. Renew Energ 2007;32:1637−44.

[138] Koyun A, ÜÇgül I, Acar M, Senol R. Bacası sisteminin termal özet Dizayni. Tesisat Mühendisligi Dergisi 2007;98:45−50.

[139] Ferreira AG, Maia CB, Cortez MFB, Valle RM. Technical feasibility assessment of a solar chimney for food drying. Sol Energy 2008;82:198−205.

[140] Ferreira AG, Goncalves LM, Maia CB. Solar drying of a solid waste from steel wire industry. Appl Therm Eng 2014;73:104−10.

[141] Maia C, Ferreira A, Valle R, Cortez M. Analysis of the airflow in a prototype of a solar chimney dryer. Heat Trans Eng 2009;30:393−9.

[142] Maia CB, Ferreira AG, Valle RM, Cortez MFB. Theoretical evaluation of the influence of geometric parameters and materials on the behavior of the airflow in a solar chimney. Comput Fluids 2009;38:625−36.

[143] Maia CB, Silva JOC, Cabezas-Gomez L, Hanriot SM, Ferreira AG. Energy and exergy analysis of the airflow inside a solar chimney. Renew Sust Energ Rev 2013;27:350−61.

[144] Maia CB, Ferreira AG, Hanriot SM. Evaluation of a tracking flat-plate solar collector in Brazil. Appl Therm Eng 2014;73:953−62.

[145] Kasaeian AB, Heidari E, Vatan SN. Experimental investigation of climatic effects on the efficiency of a solar chimney pilot power plant. Renew Sust Energ Rev 2011;15:5202–6.

[146] Kalash S, Naimeh W, Ajib S. Experimental investigation of the solar collector temperature field of a sloped solar updraft power plant prototype. Sol Energy 2013;98:70–7.

[147] Kalash S, Naimeh W, Ajib S. Experimental investigation of a pilot sloped solar updraft power plant prototype performance throughout a year. Enrgy Proced 2014;50:627–33.

[148] Wei YL, Wu ZK. Shed absorbability and tower structure characteristics of the solar heated wind updraft tower power. In: 3rd international conference on solar updraft tower technology, huazhong university of science and technology, Wuhan, China; 2012. p. 1–12.

[149] Gannon AJ, von Backström TW. Solar chimney cycle analysis with system loss and solar collector performance. J Sol Energ-T Asme 2000;122:133–7.

[150] Schlaich J, Bergermann R, Schiel W, Weinrebe G. Design of commercial solar updraft tower systems – utilization of solar induced convective flows for power generation. J Sol Energ-T Asme 2005;127:117–24.

[151] Louis T. Optimizing collector efficiency of a solar chimney power plant. In: Proceedings of melecon '85: Mediterranean electrochemical conference, solar energy; 1985. p. 219–22.

[152] Mullett LB. The solar chimney-overall efficiency, design and performance. Int J Ambient Energy 1987;8:35–40.

[153] Michaud LM. Thermodynamic cycle of the atmospheric upward heat convection process. Meteorol Atmos Phys 2000;72:29–46.

[154] Ninic N. Available energy of the air in solar chimneys and the possibility of its ground-level concentration. Sol Energy 2006;80:804–11.

[155] Lodhi MAK, Sulaiman MY. Helio-aero-gravity electric power production at low cost. Renew Energ 1992;2:183–9.

[156] Sathyajith M, Geetha SP, Ganesh B, Jeeja CK. Helio-aero-gravity effect. Appl Energ 1995;51:87–91.

[157] Lodhi MAK. Application of helio-aero-gravity concept in producing energy and suppressing pollution. Energ Convers Manage 1999;40:407–21.

[158] Pretorius JP, Kroger DG. Solar chimney power plant performance. J Sol Energ-T Asme 2006;128:302–11.

[159] von Backström TW, Fluri TP. Maximum fluid power condition in solar chimney power plants – an analytical approach. Sol Energy 2006;80:1417–23.

[160] Kröger DG, Blaine D. Analysis of the driving potential of a solar chimney power plant. SAIMechE R&D J 1999;15:85–94.

[161] Jacobs EW, Lasier DD. A theoretical analysis of solar-driven natural convection energy conversion system. In: Report DE84004431, solar energy research institute. Golden, Colorado; 1984. p. 1–27.

[162] Lodhi MAK. Thermal collection and storage of solar energies. In: Mufti A, Veziroglu TN, editors. Proceedings of international symposium: workshop on sili-contechnology development; 1987.

[163] Bouchair A, Fitzgerald D. The optimum azimuth for a solar chimney in hot climates. Energ Build 1988;12:135−40.

[164] Padki MM, Sherif SA. Fluid dynamics of solar chimneys. In: Forum on industrial applications of fluid mechanics; 1988.

[165] Padki MM, Sherif SA. Fluid dynamics of solar chimneys, In: Proceedings of ASME winter annual meeting, Chicago, IL; 1988, p. 43−6.

[166] Bernardes MAD, Valle RM, Cortez MFB. Numerical analysis of natural laminar convection in a radial solar heater. Int J Therm Sci 1999;38:42−50.

[167] Bernardes MAD, Voss A, Weinrebe G. Thermal and technical analyses of solar chimneys. Sol Energy 2003;75:511−24.

[168] Serag-Eldin MA. Mitigating adverse wind effects on flow in solar chimney plants. In: Proceedings of 4th IEC, Mansoura international engineering conference, Sharm ElSheikh; 2004. p. 20−2.

[169] Pastohr H, Kornadt O, Gurlebeck K. Numerical and analytical calculations of the temperature and flow field in the upwind power plant. Int J Energ Res 2004;28: 495−510.

[170] Pretorius JP, Kroger DG. Critical evaluation of solar chimney power plant performance. Sol Energy 2006;80:535−44.

[171] Bilgen E, Rheault J. Solar chimney power plants for high latitudes. Sol Energy 2005;79:449−58.

[172] Coetzee H. Design of a solar chimney to generate electricity employing a convergent nozzle. In: Botswana technology centre, private bag 0082, Gaborone, Botswana; 1999, p. 1−14.

[173] von Backström TW, Bernhardt A, Gannon AJ. Pressure drop in solar power plant chimneys. J Sol Energ-T Asme 2003;125:165−9.

[174] von Backström TW, Gannon AJ. Compressible flow through solar power plant chimneys. J Sol Energ-T Asme 2000;122:138−45.

[175] Kustrin I, Tuma M. Soncni dimnik. Strojniski Vestnik 1985;31:309−14.

[176] Bernardes MAD, von Backström TW. Evaluation of operational control strategies applicable to solar chimney power plants. Sol Energy 2010;84:277−88.

[177] Fluri TP, von Backström TW. Comparison of modelling approaches and layouts for solar chimney turbines. Sol Energy 2008;82:239−46.

[178] Gannon AJ, von Backström TW. Controlling and maximizing solar chimney power output. In: 1st international conference on heat transfer, fluid mechanics and thermodynamics, Kruger Park, South Africa; 2002.

[179] Gannon AJ, von Backström TW. Solar chimney turbine performance. J Sol Energ-T Asme 2003;125:101−6.

[180] Kirstein CF, von Backström TW. Flow through a solar chimney power plant collector-to-chimney transition section. J Sol Energ-T Asme 2006;128:312−17.

[181] von Backström TW, Gannon AJ. Solar chimney turbine characteristics. Sol Energy 2004;76:235−41.

[182] Stinnes WW. Extension of the feasibility study for the greenhouse operation: the 200 MW solar power station in the Northern Cape Province. In: Energy management news; 1998, p. 4−12.

[183] Dai YJ, Huang HB, Wang RZ. Case study of solar chimney power plants in Northwestern regions of China. Renew Energ 2003;28:1295−304.

[184] Onyango FN, Ochieng RM. The potential of solar chimney for application in rural areas of developing countries. Fuel 2006;85:2561−6.

[185] Beerbaum S, Weinrebe G. Solar thermal power generation in India − a techno-economic analysis. Renew Energ 2000;21:153−74.

[186] Bernardes MAS. Technical economical and ecological analysis of the solar chimney power plant systems. In: Universitat Sttgart; 2004.

[187] Michaud LM. Vortex process for capturing mechanical energy during upward heat-convection in the atmosphere. Appl Energ 1999;62:241−51.

[188] Papageorgiou CD. External wind effects on floating solar chimney. In: Proceedings of the fourth IASTED international conference on power and energy systems; 2004. p. 159−63.

[189] Papageorgiou CD. Floating solar chimney power stations with thermal storage. In: Proceedings of the sixth IASTED international conference on European power and energy systems; 2006. p. 325−31.

[190] Papageorgiou CD. Floating solar chimney technology: a solar proposal for China. In: Proceedings of ISES solar world congress 2007: solar energy and human settlement, vols. I−V; 2007. p. 172−6.

[191] Papageorgiou CD. Floating solar chimney technology for desertec. Energy Environ Eng S 2008:216−22.

[192] Papageorgiou CD, Katopodis P. A modular solar collector for desert floating solar chimney technology. Energy Environ Eng S 2009:126−32.

[193] Papageorgiou CD. Floating solar chimney technology. In: Rugescu RD, editor. Sol Energy, InTech, <http://www.intechopen.com/books/solar-energy/floatingsolar-chimney-technology>; 2010.

[194] Serag-Eldin MA. Analysis of effect of geometric parameters on performance of solar chimney plants, HT2005. In: Proceedings of the ASME summer heat transfer conference 2005, vol. 3; 2005. p. 587−95.

[195] Serag-Eldin MA. Analysis of a new solar chimney plant design for mountainous regions. Wit Trans Eng Sci 2006;53:437−46.

[196] Serag-Eldin MA. Mitigating adverse wind effects on flow in solar chimney plants. In: ASME heat transfer/fluids engineering summer conference, Charlotte, NC; 2004.

[197] Zhou XP, Yuan S, Bernardes MAD. Sloped-collector solar updraft tower power plant performance. Int J Heat Mass Tran 2013;66:798−807.

[198] Cao F, Zhao L, Guo LJ. Simulation of a sloped solar chimney power plant in Lanzhou. Energ Convers Manage 2011;52:2360−6.

[199] Cao F, Li HS, Zhang Y, Zhao L. Numerical simulation and comparison of conventional and sloped solar chimney power plants: the case for Lanzhou. Sci World J 2013.

[200] Cao F, Zhao L, Li HS, Guo LJ. Performance analysis of conventional and sloped solar chimney power plants in China. Appl Therm Eng 2013;50:582−92.

[201] Koonsrisuk A. Mathematical modeling of sloped solar chimney power plants. Energy 2012;47:582−9.

[202] Koonsrisuk A. Comparison of conventional solar chimney power plants and sloped solar chimney power plants using second law analysis. Sol Energy 2013;98:78−84.

[203] Kalash S, Naimeh W, Ajib S. A review of sloped solar updraft power technology. Enrgy Proced 2014;50:222−8.

[204] Li QL, Fan XY, Xin X, Chao J, Zhou Y. Performance study of solar chimney power plant system with a sloped collector. Key Eng Mater 2013;561:597−603.

[205] Li QL, Xie XQ, Chao J, Xin X, Zhou Y. The comparison of CFD flow field between slope solar energy power plant and traditional solar chimney power generating equipment. Key Eng Mater 2013;561:614−19.

[206] Kershman SA, Rheinlander H, Gabler H. Seawater reverse osmosis powered from renewable energy sources − hybrid wind/photovoltaic/grid power supply for small-scale desalination in Libya. Desalination 2003;153:17−23.

[207] Liu XJ, Kiyoshi T, Takeda M. Simulation of a seawater MHD power generation system. Cryogenics 2006;46:362−6.

[208] Trieb F, Muller-Steinhagen H. Concentrating solar power for seawater desalination in the Middle East and North Africa. Desalination 2008;220:165−83.

[209] Zhou XP, Xiao B, Liu WC, Guo XJ, Yang JK, Fan J. Comparison of classical solar chimney power system and combined solar chimney system for power generation and seawater desalination. Desalination 2010;250:249−56.

[210] Zuo L, Yuan Y, Li ZJ, Zheng Y. Experimental research on solar chimneys integrated with seawater desalination under practical weather condition. Desalination 2012;298:22−33.

[211] Li C, Kosmadakis G, Manolakos D, Stefanakos E, Papadakis G, Goswami DY. Performance investigation of concentrating solar collectors coupled with a transcritical organic Rankine cycle for power and seawater desalination co-generation. Desalination 2013;318:107−17.

[212] Niroomand N, Amidpour M. New combination of solar chimney for power generation and seawater desalination. Desalin Water Treat 2013;51:7401−11.

[213] Rao PS. Large scale solar power stations in India. In: IABSE congress report, <http://retro.seals.ch/digbib/view?pid = bse-cr-002>; 1992:14:167, 1992, p. 1−7.

[214] Enviromission. Technology overview, <http://www.enviromission.com.au/EVM/content/technology_technologyover.html>; 2015.

Thermodynamic fundamentals*

2

Tingzhen Ming[1,2], Yong Zheng[2], Wei Liu[2] and Yuan Pan[3]

[1]School of Civil Engineering and Architecture, Wuhan University of Technology, Wuhan, P.R. China [2]School of Energy and Power Engineering, Huazhong University of Science and Technology, Wuhan, P.R. China [3]School of Electrical and Electric Engineering, Huazhong University of Science and Technology, Wuhan, P.R. China

CHAPTER OUTLINE

2.1 INTRODUCTION

The solar chimney power plant system (SCPPS), which has the following advantages compared with the traditional power generation systems: easier to design, more convenient to draw materials, lower cost of power generation, higher

---

*Part of this chapter was published in Journal of Energy Institute.

Solar Chimney Power Plant Generating Technology. DOI: http://dx.doi.org/10.1016/B978-0-12-805370-6.00002-8

operational reliability, fewer running components, more convenient maintenance and overhaul, lower maintenance expense, no environmental contamination, continuous stable running, longer operational lifespan, is a late-model solar power generation system [1]. No related experimental results on large-scale commercial SCPPS have ever been reported since the first SC prototype was built in Spain in the 1980s, which is mainly because of the excessive early cost required [2,3]. Establishing a large-scale commercial SCPPS of over 100 MW output power requires the financial support of both local government and enterprise. In 1985, Kulunk [4] set up a miniature SCPPS experimental facility. In 1998, Pasumarthi and Sherif [5,6] built three SC models by modifying the shape and radius of the collector or canopy in Gainesville of Florida University, and carried out experiments on the temperature and velocity distributions of the airflow inside the canopy, whose results agree well with the theoretical analysis. Zhou et al. [7–9] presented some experimental and numerical results of a pilot SCPPS.

Some researchers have focused on the thermodynamic analysis of SCPPS. Michaud [10] investigated how updraft and sounding properties affect the work produced in SCPPS when air is raised and shows that the work can be transferred downward. In his work, closed ideal thermodynamic cycles were used to analyze the atmospheric upward heat convection process which was compared to the Brayton gas-turbine cycle. The heat to work conversion efficiency of the atmosphere was shown to be close to the Carnot efficiency calculated using the average temperatures at which heat is received and given up for hot and cold source temperatures, respectively. In addition, the heat to work conversion efficiency is independent of whether the lifting process is discontinuous or continuous, and nearly independent of whether the heat is transported as sensible or as latent heat.

Petela [11] thought that thermodynamic interpretation of processes occurring in these SCPPS components is based on the derived energy and exergy balances. The author presented examples of the energy and exergy flow diagrams showing how the SCPPS input of 36.81 MW energy of solar radiation, corresponding to 32.41 MW input of radiation exergy, is distributed between the SCPPS components. And responsive trends to the varying input parameters were studied. Later, the concept of mechanical exergy (ezergy) of air was applied which allowed for quantitative determination of the effect attributed to the terrestrial gravity field on the component processes of the SCPPS. The utilization of low-temperature waste heat is attracting more and more attention due to the serious energy crisis nowadays. Chen et al. [12] proposed a low-temperature waste heat recovery system based on the concept of the solar chimney. In the system, low-temperature waste heat was used to heat air to produce an air updraft in the chimney tower. The heat source temperature, ambient air temperature, and area of heat transfer were examined to evaluate their effects on the system performance, such as velocity of updraft, mass flow rate of air, power output, conversion efficiency, and exergy efficiency.

Gannon and von Backström [13] carried out a detailed thermodynamic analysis on the thermal cycle and efficiency of the whole system. In their work, they

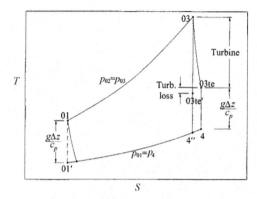

FIGURE 2.1

Temperature-entropy diagram for air standard cycle analysis with system losses for limiting turbine temperature drop [13].

presented an ideal air standard cycle analysis of the solar chimney power plant with the limiting performance, ideal efficiencies, and relationships between main variables. They also analyzed chimney friction, the system, the turbine, and exit kinetic energy losses during operation (Fig. 2.1). A simple model of the solar collector was used to include the coupling of the mass low and temperature rise in the solar collector. The method was used to predict the performance and operating range of a large-scale plant. In addition, the solar chimney model was verified by comparing the simulation of a small-scale plant with experimental data.

Ninic [14,15] presented an analysis on the available work potential of the SCPPS that atmospheric air acquires while passing through the collector. In this research, the dependence of the work potential on the air flowing into the air collector from the heat gained inside the collector, as well as air humidity and atmospheric pressure as a function of elevation were determined; various collector types using dry and humid air were analyzed; the influence of various chimney heights on the air work potential were established; and the possibly higher utilization factors of the available hot air work potential without the use of high solid chimneys were discussed. The results indicated that the vortex motion flowing downstream of the turbine can be maintained under pressure and can possibly take over the role of the solid structure chimney; a part of the available energy potential acquired in the collector would be used to maintain the vortex flow in the air column above the ground-level turbine (Fig. 2.2).

Later, Nizetic and Ninic [16,17] used working potential in turbines at ground level and analyzed the influence of different air state changes in the collector with an emphasis on overall SCPPS efficiency (Fig. 2.3). They performed an analysis for conventional SCPPS for different air temperature increases and also the influence of adding water in the collector. The results revealed that chimney efficiency in the SCPPS can be derived as a special case of the more general theory.

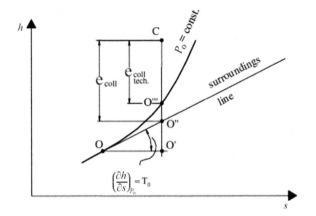

FIGURE 2.2

Exergy and its technically feasible part [14].

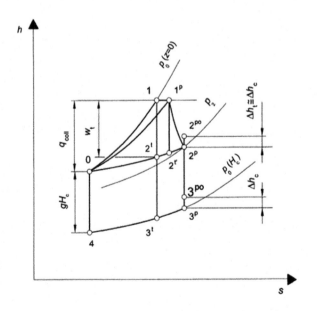

FIGURE 2.3

Theoretical and real process of air in SC power plant [17].

2.2 **THERMODYNAMIC CYCLE**

The schematic drawing of the thermodynamic process of an SCPPS is shown in Fig. 2.4. The significant state points of air flow are as follows:

1. the state of the collector inlet;
2. the state of the collector outlet, which is also the state of turbine inlet;
3. the state of the chimney inlet, which is also the state of the turbine outlet;
4. the state of the chimney outlet;
5. the state of the environment at the same height of the chimney outlet.

Analysis on the thermodynamic process inside the regions, such as the collector, the turbine, the chimney, and the environment, can be referred to [13].

Starting at the inlet of the collector, the working fluid sequentially flows through the turbine, then the chimney, and finally it releases energy into the environment, and again flows back to the collector inlet. Fig. 2.5 shows the ideal standard temperature-entropy diagram for air in SCPPS, including all systematic loss except for the negligible macroscopic kinetic energy of the chimney outlet. The thermodynamic cycle of the working fluid can be, based on the analysis in [14], simplified into the following four basic thermodynamic processes: (1) process 1-2

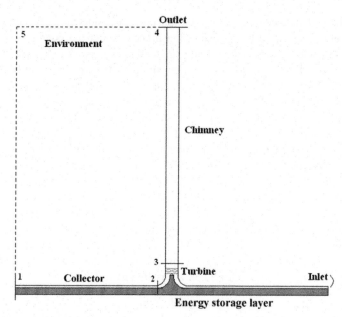

FIGURE 2.4

Schematic drawing of the thermodynamic process of SCPPSs.

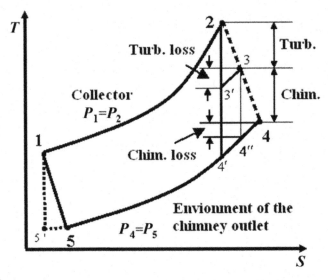

FIGURE 2.5

Temperature-entropy diagram for air standard cycle in SCPPSs.

refers to the constant-pressure heat addition process inside the collector; (2) process 2-4′ refers to the isentropic expansion process under the condition that shaft power is output from the turbine while no shaft power is output from the chimney; (3) process 4′-5 refers to the constant-pressure heat rejection process during which air flows from the chimney outlet to the environmental upper air; and (4) process 5-1 refers to the isentropic compression process during which the air flows from the environmental upper air to the collector inlet. Among the processes mentioned above, process 4′-5 and 5-1 both take place in the environment, and the above four processes compose a closed thermodynamic cycle.

The thermodynamic cycle of the SCPPS is a typical Brayton cycle, but there is an apparent difference between its network and that of a conventional Brayton cycle. The SCPPS ideal cycle 123′4′51 is a reversible cycle without irreversible energy loss. The heat absorbed during process 1-2 occurs in the collector and is the total heat input of the cycle; the output power curve is process 2-3′, but the energy consumed curve is process 3′-4′; and process 5-1, which occurs in the environment, will not consume any technical work from the system. The shaft work by the turbine for the SCPPS can be written as follows:

$$w_{shaft,i} = h_2 - h_{3'} \tag{2.1}$$

The energy that is used by the increase of potential energy of the working fluid when flowing through the chimney is:

$$w_{p,i} = h_{3'} - h_{4'} \tag{2.2}$$

As is analyzed above, process 5-1 occurs in the environment, which will not cost any technical work from the system. But the energy consumed by process 3'-4', which does come from the technical work of the system, will be approximately equal to that by process 5-1. Thereby, most of the technical work in the ideal cycle 123'4'51 is used by the increase of potential energy of the working fluid when flowing through the chimney, leaving a mini proportion to be output through shaft work by the turbine installed at the bottom of the chimney. The cycle 123451 represents the actual irreversible Brayton cycle for the SCPPS, including the turbine loss and chimney loss. The shaft work by the turbine for the SCPPS can be written as follows:

$$w_{shaft} = h_2 - h_3 = c_P T_2 \left[1 - \left(\frac{p_3}{p_2} \right)^{\frac{k-1}{k}} \right] \tag{2.3}$$

The energy used by the increase of potential energy of the working fluid when flowing through the chimney is:

$$w_p = h_3 - h_4 \tag{2.4}$$

Compared with the results shown in Eqs. (2.3) and (2.4), the value of w_p will be much higher than that of w_{shaft}.

2.3 THERMAL EFFICIENCY

Under the condition of steady solar radiation during daytime, the thermal efficiency of the cycle η can be written as follows:

$$\eta = \frac{w_{shaft}}{q_{12}} \tag{2.5}$$

where, q_{12} can be written as:

$$q_{12} = h_2 - h_1 = c_P(T_2 - T_1) \tag{2.6}$$

Substituting Eqs. (2.3) and (2.6) into Eq. (2.5), we have:

$$\eta = \frac{T_2}{T_2 - T_1} \left[1 - \left(\frac{p_3}{p_2} \right)^{\frac{k-1}{k}} \right] = \frac{(T_1 + \Delta T)}{\Delta T} \left[1 - \left(1 - \frac{\Delta p_{turb}}{p_1} \right)^{\frac{k-1}{k}} \right] \tag{2.7}$$

In the equation above, Δp_{turb} stands for the pressure drop across the turbine: $\Delta p_{turb} = P_2 - P_3 = P_1 - P_3$, while ΔT represents the temperature rise of working fluid inside the collector: $\Delta T = T_2 - T_1$. In Eq. (2.7), the value of ΔT can be provided explicitly by numerical simulations, which will be strongly related to the following parameters: solar radiation, turbine pressure drop, etc. Thereby, the actual efficiency of the SCPPS will be strongly related to the actual operation process which is controlled by various parameters: solar radiation, turbine pressure drop, ambient temperature, geometric dimensions of the SCPPS, etc.

The ideal thermal efficiency of the solar chimney cycle, if all irreversible losses are neglected, can be expressed as follows:

$$\eta_i = \frac{q_{12} - |q_{4'5'}|}{q_{12}} = \frac{\Delta h_{24'} - \Delta h_{3'4'}}{q_{12}} \tag{2.8}$$

thus:

$$\eta_i = \frac{(h_2 - h_{4'}) - (h_{1'} - h_{5'})}{h_2 - h_1} = \frac{(T_2 - T_{4'}) - (T_{1'} - T_{5'})}{T_2 - T_1} \tag{2.9}$$

According to the characteristics of each process in the Brayton cycle, we can get:

$$\frac{T_1}{T_{5'}} = \frac{T_2}{T_{4'}} = \left(\frac{p_1}{p_{5'}}\right)^{\frac{\kappa-1}{\kappa}} = \left(\frac{p_2}{p_{4'}}\right)^{\frac{\kappa-1}{\kappa}} = \pi^{\frac{\kappa-1}{\kappa}} \tag{2.10}$$

where, π is the pressure ratio: $\pi = \frac{p_1}{p_{5'}}$. Substituting Eq. (2.10) into Eq. (2.9), we have:

$$\eta_i = 1 - \frac{T_{5'}}{T_1} = 1 - \frac{1}{\pi^{\frac{\kappa-1}{\kappa}}} \tag{2.11}$$

So the ideal efficiency of the SCPPS is exactly equal to that of the conventional Brayton cycle. When the hot air flows from the bottom through the chimney to the environment, or the same mass of cold air comes from the environment at the same height of chimney outlet to the collector inlet, the energy transfer can be written as:

$$c_P dT = gdz$$

By integrating this equation, we have:

$$c_P(T_1 - T_{5'}) = gH \tag{2.12}$$

From Eqs. (2.12) and (2.11), we can obtain:

$$\eta_i = 1 - \frac{1}{\pi^{\frac{\kappa-1}{\kappa}}} = \frac{gH}{c_P T_1} \tag{2.13}$$

Thereby, the ideal efficiency of the SCPPS is related to the chimney height and the ambient temperature, if we neglect the difference between the process 5-1, process 5'-1, and process 3'-4'. The analysis result agrees well with that shown in Ref. [13]. This efficiency in Eq. (2.13) cannot be obtained because many influencing factors are not taken into account.

2.4 RESULTS AND ANALYSIS

To validate the theoretical analysis and to compare with the results between the ideal efficiency in Eq. (2.13) and the actual efficiency in Eq. (2.7) shown above,

Table 2.1 Geometric Dimensions of Three Typical SCPPSs

Model	Chimney		Collector	
	Height (m)	Diameter (m)	Height (m)	Diameter (m)
Spanish prototype	200	10	2~6	122
MW-graded	400	50	2~10	1000
100 MW-graded	1000	130	3~25	2500

Table 2.2 Basic Parameters of Three Typical SCPPSs

Parameters	Units	Value
Transmissivity of canopy	–	0.9
Emissivity of canopy	–	0.85
Absorptivity of storage layer	–	0.9
Conductivity of storage layer	W/(m·K)	1.2
Thickness of storage layer	m	2.0
Ambient temperature	K	293.15
Ambient air velocity	m/s	2
Solar radiation intensity	W/m^2	500, 750, 1000
Turbine pressure drop	Pa	0~1500
Efficiency of turbine	–	72%

three typical types of SCPPs were analyzed by numerical simulation. The mathematical models to describe the flow and heat transfer characteristics can be found in Refs. [11,12], the geometric dimensions for the models can be found in Table 2.1, and the related parameters with the same values are shown in Table 2.2.

Boundary conditions are set as follows: For the roof of the collector, we take convection boundary into account, and coefficient of convection is set as 10 W/(m^2·K) which can be accepted when the environment air velocity is not very large, that is, 1−2 m/s. The temperature of the environment is set as 293 K, the inlet of the collector is set as the pressure-inlet boundary, the chimney wall can be set as an adiabatic boundary; the chimney outlet is set as the pressure-outlet boundary; the bottom of the energy storage layer is set as the temperature-constant boundary, whose temperature is 300 K. Solar radiation which projects through the transparent ceiling into the ground can be considered as a heat source for the ground thin layer [10]. We can simulate the running conditions of an SCPPS on the whole by setting solar radiation as 500, 750, and 1000 W/m^2, respectively. The acceleration of gravity is 9.8 m/s^2, and the air density changing with altitude can be found in Ref. [9]. Thereby, the pressure ratio can be calculated accordingly.

In addition, it is necessary to explain the reason for why we consider the collector inlet and chimney outlet both as pressure boundaries and have their pressures set as 0 Pa. This is because we simultaneously take the inner and outer

Table 2.3 Boundary Conditions

Place	Type	Value
0.1 mm top layer of the ground	Heat source	$2 \times 10^6 \sim 8 \times 10^6$ W/m^3
Bottom of the ground	Temperature	300 K
Surface of the canopy	Wall	$T = 293$ K, $\alpha = 10$ W/(m$^2 \cdot$K)
Surface of the chimney	Wall	$q_{chim} = 0$ W/m^2
Collector inlet	Pressure inlet	$p_{r,inlet} = 0$ Pa, $T_0 = 293$ K
Chimney outlet	Pressure outlet	$p_{r,outlet} = 0$ Pa

pressure distributions of the system into account, $p_{r,i} = 0$ means that for both the inside and outside of the collector inlet, the static pressures at the same height are the same [18,19].

However, the setting of pressure drop across the turbine in this paper differs from the processing method applied by Pastohr [20]. The turbine of SCPPS, as explained earlier, belongs to a pressure-based wind turbine, the fore-and-aft air velocities are almost the same but the pressure changes significantly, and its power output does not follow the Beetz power limit theory. Therefore, the output power through the turbine can be calculated according to Eq. (2.14) by presetting the pressure drop across the turbine:

$$W_{shaft} = \eta_{shaft} \cdot \Delta p \cdot V \qquad (2.14)$$

where, W_{shaft} represents the shaft power output through the turbine, η_{shaft} represents the energy conversion efficiency of the turbine, which can be preset as 80% (less than the optimized data), Δp represents the pressure drop across the turbine, V represents the air volume flow rate of the system flowing through the chimney outlet. The boundary conditions for different places are shown in Table 2.3.

The standard $k - \varepsilon$ model is applied during the numerical simulation of air flow in the collector and chimney, the SIMPLE algorithm is applied for pressure-velocity coupling, and the momentum equation, energy equation, and other equations all apply the second-order upwind discretization scheme. The mesh numbers of the 50 kW, MW-graded, and 100 MW-graded SCPPSs are nearly 500,000, 1,200,000, and 2,500,000, respectively, where we can get grid-independent simulation results.

2.4.1 COMPUTATION RESULTS FOR THE SPANISH PROTOTYPE

Figs. 2.6 and 2.7 show the numerical results of the SC prototype in Spain. As shown in Fig. 2.6, when the solar radiation intensity is 1000 W/m^2, the maximum output power is about 75 kW—about 50% higher than the design value of the Spanish prototype. This is because the efficiency of turbine shown in Table 2.2 is almost 50% higher than the design value which is based on the Betz theory.

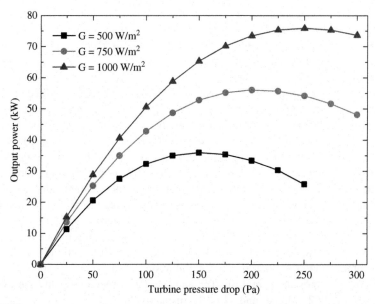

FIGURE 2.6

Output power of the SC prototype in Spain.

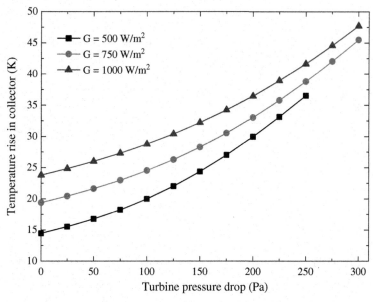

FIGURE 2.7

Temperature rise inside the collector.

The efficiency of a conventional turbine, which is used in free wind farms, with a good design based on the Betz theory is a little less than 50%, while the turbines used in the solar chimney systems are pressure-staged turbines and their efficiencies will be higher than 85%. Thereby, the efficiency of the turbine selected in Table 2.2 is acceptable and reasonable.

Fig. 2.7 shows the temperature rise of the airflow inside the collector. It can be easily seen that the air temperature rise inside the collector increases significantly with the increase of solar radiation intensity and turbine pressure drop. Apparently, the total pressure drop of the SCPPS is P_1-P_5, which is exactly determined by the chimney height and the air density distribution along the altitude of the ambience. With the increase of pressure drop across the turbine, the pressure drop by harnessing the air flow through the chimney decreases, and the air flow rate and velocity will decrease. Therefore, the air temperature rise inside the collector will increase. In addition, the air temperature rise inside the collector also increases with the increase of solar radiation intensity. From Figs. 2.6 and 2.7, we can see that the turbine pressure drops according to the maximum output power: when the solar radiation intensities are 500, 750, and 1000 W/m^2, the pressures are 150, 200, and 250 Pa, respectively, and the temperature rises are 28.28, 33.04, and 41.65 K, respectively.

Fig. 2.8 shows the effects of pressure drop across the turbine on the cycle thermal efficiency, in which the efficiency curve of the ideal value refers to the

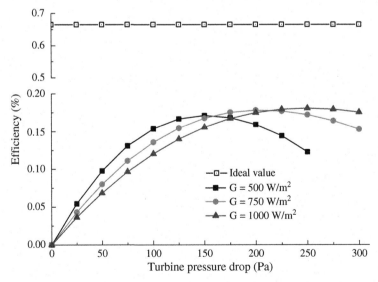

FIGURE 2.8

Cycle efficiencies of SC prototype in Spain.

efficiency of Brayton cycle shown in Eq. (2.13). From this figure, we can find that the ideal value of the efficiency of the Spanish solar chimney power plant prototype is 0.665%, while the actual efficiencies corresponding to the changing turbine pressure drop of the prototype are all less than 0.2% under different solar radiation intensities. The ideal thermal efficiency of the system is lower than 1%, for which the main reason is that most of the heat energy from the solar energy cannot transfer into shaft work during the isentropic expansion process in the turbine (turbines), it can just be used to overcome gravity when flowing through the chimney.

Moreover, the ideal thermal efficiency of the system is much higher than the actual efficiencies, this is because the increase of the turbine pressure drop results in a larger temperature rise in the collector shown in Fig. 2.7, accompanied by a larger heat loss through the collector canopy and a larger energy loss through the chimney outlet. Therefore, the SCPPS is unable to reach the ideal thermal efficiency level of the standard Brayton cycle shown in Fig. 2.8. In addition, the actual efficiency of the system is strongly related to the geometric dimensions, the turbine pressure drop, the solar radiation intensity, and all the parameters shown in Table 2.2, while the ideal efficiency is only subject to the chimney height and ambient temperature. Thereby, the actual efficiency of the SCPPS can never reach the value of ideal efficiency, and the former will be more useful for the design and commercial application of different kinds of SCPPSs.

2.4.2 COMPUTATION RESULTS FOR COMMERCIAL SCPPSs

Now we carry out thermodynamic analysis on an MW-graded solar chimney whose geometric dimensions are shown in Table 2.1, computing for its output power and cycle thermal efficiency. Fig. 2.9 shows the comparison relationship among the efficiencies of actual cycle and the ideal cycle of the MW-graded SCPPS. As shown in this figure, the MW-graded system ideal cycle efficiency is about 1.33%, which is independent of all the actual operational parameters, such as solar radiation intensity and turbine pressure drop. The maximum value of the system actual efficiencies is near 0.3%, which is lower than one quarter of the ideal efficiency value. By comparison, the difference between the actual and ideal efficiencies becomes larger with the dimensions of the solar chimney system, this is because the increase of collector area will decrease the system efficiency due to a larger heat energy loss through the canopy.

The geometric dimensions of large-scale SCPPS are shown in Table 2.1, and the simulation results are shown in Figs. 2.10 and 2.11. As shown in Fig. 2.10, the ideal efficiency of a large-scale SCPPS with chimney height 1000 m is about 3.33%, while the maximum value of the actual efficiencies is about 0.9%. Obviously, the actual efficiency is also much lower than the ideal value, which is also because of much larger heat energy loss from the collector canopy.

Although the actual efficiency of the large-scale SCPPS is not higher than 0.9%, the maximum output power is over 100 MW when the solar radiation

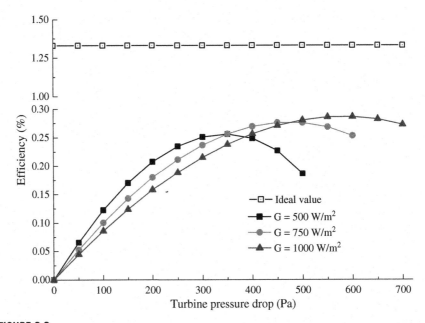

FIGURE 2.9

Cycle efficiencies of MW-graded SCPPS.

intensity is 750 W/m^2 or more. Taking into consideration that the SCPP will not consume any fossil energy, the raw material used to build the system is easy to access, and the investment for the system is relatively low, the actual efficiency, which is about 1%, is acceptable for its commercial application.

Further study will focus on the following two points: (1) quantitatively explore the detailed factors which have effect on the actual efficiency; (2) find a new way to increase the total efficiency and output power of the system, for instance: combining the SCPP and the photovoltaic power system to build a new synthetic system to highly improve thermal efficiency of the system.

2.5 EFFECT OF VARIOUS PARAMETERS

As we know, it is very difficult to quantitatively present a theoretical analysis of the influences of various parameters on the efficiency of Solar Updraft Power Plant System (SUPPS) due to the strong intercorrelations of these parameters, so numerical analysis on the SUPPS by varying one parameter with the other parameters being constant becomes an effective way to solve this problem. The key parameters influencing the system efficiency are turbine efficiency, system geometrical dimensions, solar radiation, and ambient temperature. These will be discussed one by one.

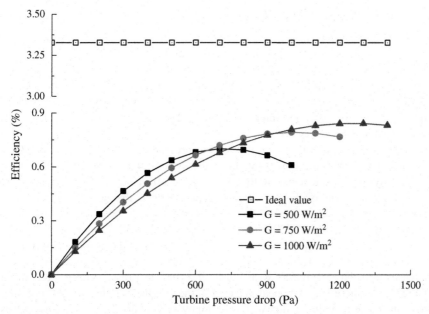

FIGURE 2.10

Cycle efficiencies of large-scale SCPPS.

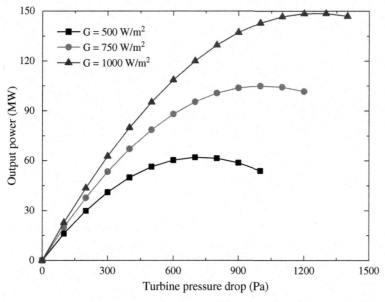

FIGURE 2.11

Output power of large-scale SCPPS.

2.5.1 INFLUENCE OF TURBINE EFFICIENCY

When conducting our simulations to study the influence of turbine efficiency on the system efficiency, the SUPPS structural parameters were selected as follows: the diameter of the collector 2000 m and the height and diameter of the chimney 900 and 90 m, respectively. The solar radiation value was selected as 500 W/m^2 and the ambient temperature 20°C. The turbine efficiency was allowed to vary from 0.72 to 0.9, with an interval of 0.06. Simulation results are shown in Figs. 2.12 and 2.13.

According to Fig. 2.12, when the turbine pressure drop is higher than 200 Pa and in a large range, the power output of the system could be over 10 MW with different system efficiencies. Setting the turbine efficiency to 90%, the highest system output power is close to 20 MW. This suggests that it is not too hard to achieve an output power of 10 MW under normal conditions with general solar radiation in the vicinity of 500 W/m^2 for this physical SUPPS model.

Fig. 2.13 shows the influence of turbine efficiency and turbine pressure drop on system efficiency. Under certain turbine efficiencies, the system efficiency increases and then decreases with increasing turbine pressure drop. According to Fig. 2.12, there exists a peak value for the product of the turbine pressure drop and outflow volume flux at the chimney outlet though the flux decreases with increasing pressure drop. Thus the system power output and efficiency also have a peak value. Considering that the cost of the turbine is much smaller than

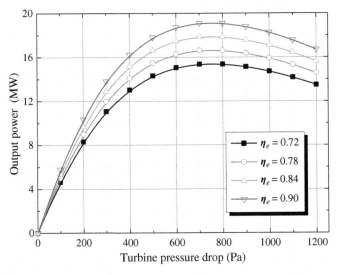

FIGURE 2.12

Influence of turbine efficiency and pressure drop on system output power.

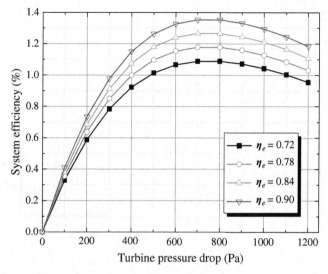

FIGURE 2.13

Influence of turbine efficiency and pressure drop on system efficiency.

that of the collector and the chimney in a given SUPPS, improving the efficiency of the turbine is probably the most effective method to improve the overall system efficiency.

2.5.2 INFLUENCE OF CHIMNEY HEIGHT AND DIAMETER

For the simulations intended to examine the influence of chimney height and diameter on the system efficiency, the height of the chimney was allowed to range from 800 to 1200 m while the diameter varied from 40 to 140 m. The diameter of the collector was set to 2000 m. Solar radiation was selected as 500 W/m^2 while the ambient temperature was set at 20°C. The turbine pressure drop and efficiency were set at 800 Pa and 0.72, respectively. Simulation results are shown in Fig. 2.14.

In Fig. 2.14, for a certain chimney height, the system efficiency increases with the increase of the chimney diameter, however the increase of the system efficiency tapers off at some point and we reach a state of diminishing return. Thus, it would probably be advisable to increase the chimney diameter only up to a point (around 100 m) as a way of increasing the system efficiency, and refrain from doing so beyond chimney diameters of 100 m. The effect of chimney height, on the other hand, plays a bit of a different role when it comes to system efficiency. For example, for a chimney diameter of 100 m, the system efficiency increases from 0.98% to 1.4% when the chimney height is increased from 800 to 1200 m. Thus the height seems to be a key parameter to use to increase system

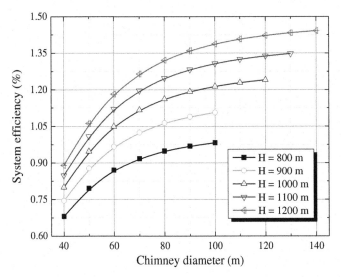

FIGURE 2.14

Influence of chimney height and diameter on system efficiency.

efficiency. However, proper chimney height should be chosen taking into account cost issues as an increase in height necessarily translates into an increase in cost.

2.5.3 INFLUENCE OF COLLECTOR DIAMETER

For the simulations intended to examine the influence of collector diameter on the system efficiency, the diameter of the collector ranged from 1000 to 3000 m, while the chimney height and diameter were set at 900 and 90 m, respectively. The solar radiation value used in the simulations was 500 W/m^2 as before, while the ambient temperature remained at 20°C. The turbine efficiency was set at 72%. Simulation results are shown in Fig. 2.15.

The combined effect of the collector diameter and turbine pressure on the system efficiency is a bit complex. For a given collector diameter, the efficiency first increases with the turbine pressure drop and then slowly decreases. For a given turbine pressure drop, the efficiency decreases with an increase in the collector diameter. A part of the energy absorbed by the collector is used to heat the air and increase the air velocity, another part is transferred to the ground, and the rest escapes to the environment via the collector canopy. We could attempt to explain this behavior as follows. When the area of the collector is zero, the flow of air in the system would only depend on the density difference between the air inside the chimney body and the ambient air, with no losses to the ground or the environment. As the collector diameter increases, there will be energy losses to the ground and environment, with an associated reduction in the efficiency of

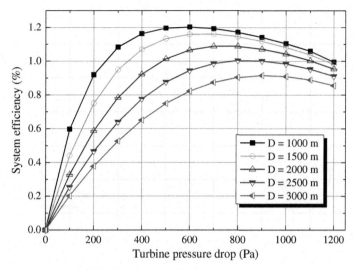

FIGURE 2.15

Influence of collector diameter and turbine pressure drop on system efficiency.

transferring the heat to kinetic energy. Furthermore, it can be seen that the turbine pressure drop for maximum system efficiency increases with the increase in collector diameter. This is attributed to the fact that the increase of the collector area causes the system draft power to increase.

2.5.4 THE INFLUENCE OF THE SOLAR RADIATION

In the simulations conducted to study the influence of solar radiation on the system efficiency, the solar radiation was varied from 400 to 800 W/m² with a 100 W/m² interval. The chimney height and diameter were kept at 900 and 90 m, respectively. The collector diameter was set at 2000 m. The ambient temperature was kept at 20°C, while the turbine efficiency was set at 72%. Results of the simulations are shown in Fig. 2.16.

As can be seen, the relationship among solar radiation, turbine pressure drop, and system efficiency is a bit complex. When the pressure drop and radiation are both low, most of the input energy is used to heat the air in the system, thus contributing to a higher efficiency. At higher values of the pressure drop, the air flow drops, while the temperatures of both the soil surface and air increase, thus contributing to transferring more energy to the ground and the environment. All of this contributes to decreasing the efficiency. When the solar radiation is higher, only by increasing the turbine pressure drop to produce enough axial work can the system efficiency become higher.

A certain system efficiency value, except the peak, has two corresponding turbine pressure drop values. Low turbine pressure drop causes an increase in the air

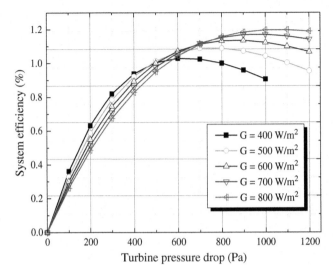

FIGURE 2.16

Influence of solar radiation and turbine pressure drop on system efficiency.

flow rate, while a higher pressure drop decreases the flow rate and thus contributes to increasing the energy stored. To make the system operate more steadily, it is better to select the higher pressure drop values.

2.5.5 THE INFLUENCE OF AMBIENT TEMPERATURE

To investigate the influence of the environmental temperature on the system efficiency, the ambient temperature was varied from $-10°C$ to $30°C$, while the collector diameter was set at 2000 m. The diameter and height of chimney were assigned the values of 90 and 900 m, respectively. The solar radiation was set at 500 W/m^2 and the turbine efficiency was set at 72%. The results are shown in Fig. 2.17.

From this figure, it can be seen that, for a certain turbine pressure drop, the system efficiency increases with a decrease in the ambient temperature. When the pressure drop is small, the system efficiency increases slightly with the decrease of the ambient temperature. When the pressure drop is large, the efficiency increases more rapidly with the decrease of the ambient temperature. It is important to design the system so that it operates under turbine pressure drop values close to those corresponding to the efficiency's peak value.

It should be mentioned, that a lower ambient temperature may not be the best choice for a SUPPS. Generally, in the winter and spring, the solar radiation is comparatively weak, so close attention should be paid to whether all parts are operating normally at lower ambient temperatures.

Overall, various parameters show complex influences on the efficiency of SUPPS. The performance of the turbine seems to be the most important to

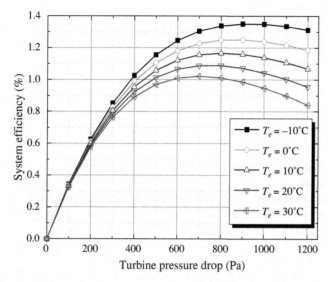

FIGURE 2.17

Influence of ambient temperature and turbine pressure drop on system efficiency.

SUPPS. Increasing the turbine efficiency seems to greatly improve the overall efficiency of SUPPS. Significantly reducing the geometrical dimensions of the system is necessarily associated with a significant reduction of the total construction cost. Higher chimney and strong solar radiation will cause larger driving forces which will greatly increase the system efficiency. For the collector diameter and ambient temperature, decreasing these two parameters will cause higher overall system efficiency. Similarly, decreasing the collector diameter will greatly decrease the amount of solar radiation energy.

2.6 CONCLUSIONS

This chapter has made an analysis of the flow of working fluid within various parts of SCPPS and a thermodynamic cycle, starting from the collector inlet, passing through the collector and chimney outlet, and finally back to the collector inlet from the environment. In addition, numerical models for ideal cycle efficiency and actual cycle efficiency are also established. Computations and predictions are carried out for SCPPSs of various scales, and the research results show that: under the same pressure drop as the Spanish solar chimney power plant prototype, the ideal efficiency of the standard Brayton cycle is about 0.665%, while its actual cycle efficiency is less than 0.2%; the ideal cycle efficiency and maximum actual efficiency of the standard Brayton cycle corresponding to medium-scale SCPPS are 1.33% and 0.3%, respectively, while the same parameters for large-scale SCPPSs are 3.33% and 0.9%. The actual efficiency of any type of SCPPS is much lower than the ideal efficiency based on the Brayton cycle, and the former can be used for the design and commercial application of large-scale SCPPSs.

A two-dimensional steady state numerical simulation on a commercial SUPPS to analyze the influence of various parameters on the system efficiency has been presented. Mathematical models to describe the fluid flow, heat transfer, and output power were investigated, employing a method to incorporate the presence of the turbine. Based on the results obtained, we can reach the following findings.

1. The turbine plays an important role in the efficiency of SUPPS. Increasing the turbine efficiency will greatly improve the overall performance of SUPPS and reduce its construction cost. The turbine pressure drop also has a significant influence on the system efficiency, output power, and the energy stored.
2. Chimney height is another key parameter having significant influence on the efficiency of SUPPS. Increasing the chimney height will greatly increase the system driving force, output power, and ultimately the efficiency. But taller chimneys will significantly increase construction costs for a commercial SUPPS. Increasing chimney diameter will be beneficial to system efficiency but will also cause construction costs to rise.
3. Solar radiation is always a positive factor to system performance, so it plays a very important role in the site selection of SUPPS.
4. Increasing the collector diameter and ambient temperature will both have negative effects on the system efficiency. However, the collector is the only solar radiation receiver and without it no output power will be generated.

NOMENCLATURE

c_p	Specific heat at constant pressure, kJ/(kg·K)
G	Solar radiation intensity, W/m^2
g	Gravity constant, m/s^2
h	Enthalpy, kJ/kg
H	Chimney height, m
P	Pressure, Pa
q	Heat absorbed by the working fluid, kJ
T	Temperature, K
W	Work, kJ

SUBSCRIPT

i	Ideal process
p	Used to increase potential energy of the air
$shaft$	Shaft

12	Thermodynamic process 12
turb	Turbine
inlet	Collector inlet
outlet	Chimney outlet
r	Relative pressure

GREEK SYMBOLS

η	THERMAL efficiency
α	Heat transfer coefficient, $W/(m^2 \cdot K)$
Δ	Difference
π	Pressure ratio
κ	Specific heat ratio

REFERENCES

[1] Schlaich J, Bergermann R, Schiel W, Weinrebe G. Design of commercial solar updraft tower systems—utilization of solar induced convective flows for power generation. J Sol Energ-T Asme 2005;127:117−24.

[2] Haaf W, Friedrich K, Mayer G, Schlaich J. Solar chimneys. Int J Solar Energy 1983;2:3−20.

[3] Haaf W, Friedrich K, Mayer G, Schlaich J. Solar chimneys. Int J Solar Energy 1984;2:141−61.

[4] Kulunk H. A prototype solar convection chimney operated under izmit condition. In: Veziroglu TN, editor. Proc. 7th MICAS, 1985, p. 162.

[5] Pasumarthi N, Sherif SA. Experimental and theoretical performance of a demonstration solar chimney model—Part I: mathematical model development. Int J Energ Res 1998;22:277−88.

[6] Pasumarthi N, Sherif SA. Experimental and theoretical performance of a demonstration solar chimney model—Part II: experimental and theoretical results and economic analysis. Int J Energ Res 1998;22:443−61.

[7] Zhou XP, Yang JK, Xiao B, Hou GX. Experimental study of temperature field in a solar chimney power setup. Appl Therm Eng 2007;27:2044−50.

[8] Zhou XP, Yang JK, Xiao B, Long F. Numerical study of solar chimney thermal power system using turbulence model. J Energy Inst 2008;81:86−91.

[9] Zhou XP, Yang JK, Xiao B, Hou GX. Simulation of a pilot solar chimney thermal power generating equipment. Renew Energ 2007;32:1637−44.

[10] Michaud LM. Thermodynamic cycle of the atmospheric upward heat convection process. Meteorol Atmos Phys 2000;72:29−46.

[11] Petela R. Thermodynamic study of a simplified model of the solar chimney power plant. Sol Energy 2009;83:94−107.

[12] Chen K, Wang JF, Dai YP, Liu YQ. Thermodynamic analysis of a low-temperature waste heat recovery system based on the concept of solar chimney. Energ Convers Manage 2014;80:78−86.

[13] Gannon AJ, von Backström TW. Solar chimney cycle analysis with system loss and solar collector performance. J Sol Energ-T Asme 2000;122:133−7.

[14] Ninic N. Available energy of the air in solar chimneys and the possibility of its ground-level concentration. Sol Energy 2006;80:804–11.

[15] Ninic N, Nizetic S. Elementary theory of stationary vortex columns for solar chimney power plants. Sol Energy 2009;83:462–76.

[16] Nizetic S, Ninic N. Analysis of overall solar chimney power plant efficiency. Strojarstvo 2007;49:233–40.

[17] Nizetic S, Ninic N, Klarin B. Analysis and feasibility of implementing solar chimney power plants in the Mediterranean region. Energy 2008;33:1680–90.

[18] Bernardes MAD, Voss A, Weinrebe G. Thermal and technical analyses of solar chimneys. Sol Energy 2003;75:511–24.

[19] Zheng Y, Ming TZ, Zhou Z, Yu XF, Wang HY, Pan Y, et al. Unsteady numerical simulation of solar chimney power plant system with energy storage layer. J Energy Inst 2010;83:86–92.

[20] Pastohr H, Kornadt O, Gurlebeck K. Numerical and analytical calculations of the temperature and flow field in the upwind power plant. Int J Energ Res 2004;28:495–510.

Helio-aero-gravity (HAG) effect of SUPPS*

3

Tingzhen Ming[1,2], Wei Liu[2] and Guoliang Xu[2]

[1]School of Civil Engineering and Architecture, Wuhan University of Technology, Wuhan, P.R. China [2]School of Energy and Power Engineering, Huazhong University of Science and Technology, Wuhan, P.R. China

CHAPTER OUTLINE

3.1 INTRODUCTION

The widespread use of solar energy, as an alternate and nondepletable resource for agriculture and industry as well as other applications, is dependent on the development of solar systems which possess the reliability, performance, and economic characteristics that compare favorably with the conventional systems. The solar chimney power plant system (SCPPS), which is composed of the solar collector, the chimney and the turbine, has been investigated all over the world since the German researcher Jorg Schliaich first made the brainchild in the 1970s. The main objective of the collector is to collect solar radiation to heat up the air inside. As the air density inside the system is less than that of the environment at the same height, natural convection affected by buoyancy, which acts as driving force, comes into existence. Due to the existence of the chimney, the cumulative buoyancy results in a large pressure difference between the system and the

---

*This chapter is adapted and expanded according to the paper published in International Journal of Energy Research.

environment, and the heated then air rises up into the chimney with great speed. If an axis-based turbine is placed inside the chimney where there is a large pressure drop, the potential and heat energy of the air can be converted into kinetic energy and ultimately into electric energy.

After Schlaich's pioneering work on the solar chimney concept to harness solar energy, Haaf et al. [1] provided the fundamental investigations for the Spanish prototype system in which the energy balance, design criteria, and cost analysis were discussed. Later, the same authors [2] reported preliminary test results of the solar chimney power plant. Krisst [3] demonstrated a "back yard type" device with a power output of 10 W in West Hartford, Connecticut, USA. Kulunk [4] produced a microscale electric power plant of 0.14 W in Izmit, Turkey. Pasumarthi and Sherif [5] developed a mathematical model to study the effect of various environment conditions and geometry on the air temperature, air velocity, and power output of the solar chimney. Pasumarthi and Sherif [6] also developed three model solar chimneys in Florida and reported the experimental data to assess the viability of the solar chimney concept. Padki and Sherif [7] developed a simple model to analyze the performance of the solar chimney. Lodhi [8] presented a comprehensive analysis of the chimney effect, power production, efficiency, and estimated the cost of the solar chimney power plant set up in developing nations. Bernardes et al. [9] presented a theoretical analysis of a solar chimney, operating on natural laminar convection in steady state. Gannon and Backström [10] presented an air standard cycle analysis of the solar chimney power plant for the calculation of limiting performance, efficiency, and relationship between main variables including chimney friction, and system, turbine, and exit kinetic energy losses. Bernardes et al. [11] developed a thermal and technical analysis to estimate the power output and examine the effect of various ambient conditions and structural dimensions on the power output. Pastohr et al. [12] carried out a numerical simulation to improve the description of the operation mode and efficiency by coupling all parts of the solar chimney power plant including the ground, collector, chimney, and turbine. Schlaich [13] presented the theory, practical experience, and economy of the solar chimney power plant to give a guide for the design of 200 MW commercial SCPPSs. Ming et al. [14] presented a thermodynamic analysis on the solar chimney power plant and advanced energy utilization degree to analyze the performance of the system. Liu et al. [15] carried out a numerical simulation for the MW-graded solar chimney power plant, presenting the influences of pressure drop across the turbine on the draft and the power output of the system.

For a SCPPS with certain geometrical dimensions, the important factors which influence the power output are solar radiation, chimney height, and volumetric flux, which have been analyzed and validated by theoretical and experiment investigations. However, the role of pressure difference on the performance of solar chimney systems has long been ignored. Many researchers only recognize the pressure difference as a function of the air density difference and the chimney height [1,8] and Bernardes et al. [11] neglected the theoretical analysis of pressure in the system but gave a comparatively simple driving force expression. Gannon and Backström [10], Krisst [3], Kulunk [4], Pasumarthi [5,6], and Padki [7] also ignored the discussion on pressure difference. Recently, Pastohr et al. [12] presented pressure profiles of the collector numerically, but in his investigation the static pressure inside the collector is positive and increases along the flow direction, which is in contradiction with basic flow theory and the solar chimney principle. Therefore, the effect of pressure difference on the performance of solar chimneys and the pressure field in the system remain unsolved problem, which motivated us to do a thorough analysis on the pressure distribution, exploring the relationship between the relative static pressure and driving force, and predicting the power output and efficiency.

3.2 RELATIVE STATIC PRESSURE

Consider the static pressure profiles inside and outside the chimney as shown in Fig. 3.1. Denoting S the static pressure difference between the chimney and the environment at the same height: $S = p_i - p_o$

Thereby, S can be regarded as the relative static pressure. Hence S at x and $x + dx$ can be written as:

$$S_x = p_{i,x} - p_{o,x} \tag{3.1}$$

$$S_{x+dx} = S_x + \frac{dS_x}{dx} dx \tag{3.2}$$

From the definition of S and the equations above, we have:

$$\frac{dS_x}{dx} = \frac{dp_{i,x}}{dx} - \frac{dp_{o,x}}{dx} \tag{3.3}$$

In the environment, the relationship between the static pressure and air density can be written as:

$$\frac{dp_{o,x}}{dx} = -\rho_{o,x} g \tag{3.4}$$

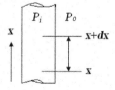

FIGURE 3.1

Static pressure inside and outside the chimney.

Considering a steady adiabatic SCPPS of cylindrical geometry and neglecting the viscous friction, we can write the momentum equation as:

$$\frac{dp_{i,x}}{dx} = -\rho_{i,x}g \tag{3.5}$$

Substituting Eqs. (3.3), (3.4), and (3.5) into Eq. (3.2) yields:

$$dS_x = S_{x+dx} - S_x = \left(\rho_{o,x} - \rho_{i,x}\right)gdx \tag{3.6}$$

Integrating Eq. (3.6) from 0 to x along the chimney yields:

$$S_x - S_0 = \int_0^x \left(\rho_{o,x} - \rho_{i,x}\right)gdx \tag{3.7}$$

At the chimney exit, the static pressure must be equal to the ambient atmospheric static pressure at that altitude. Thereby, we get:

$$S_H = 0$$

Integrating Eq. (3.7) we can get:

$$S_0 = -(S_H - S_0) = -\int_0^H \left(\rho_{o,x} - \rho_{i,x}\right)gdx \tag{3.8}$$

For the small scale SCPPSs, the density inside the chimney and the environment can be assumed to be constant. So Eq. (3.8) can be written as:

$$S_0 = -\left(\rho_o - \rho\right)gH \tag{3.9}$$

Substituting Eq. (3.9) into Eq. (3.7) yields:

$$S_x = -\left(\rho_o - \rho\right)g(H - x) \tag{3.10}$$

From Eqs. (3.9) and (3.10), we can easily find that, in the chimney, S has a negative value distribution. In addition, S has a linear relationship with the height of the chimney and the minimum value of S lies in the bottom of the chimney.

Although the expression of S is deduced in the chimney, it can also be used in the whole SCPPS. As the air is heated persistently inside the collector, the density may be less than that of the environment at the same height. Thus S also has a negative value distribution inside the collector.

In the collector, as the air velocity is not very large, the viscous resistance can be neglected. From the Continuity Equation and Bernoulli Equation, we get:

$$v_r = \frac{\dot{m}}{\rho A_r}$$

$$p_{i,r} + \rho_r gx + \frac{1}{2}\rho v_r^2 = const$$

As the height and velocity only change slightly along the flow direction inside the collector, the static pressure also changes slightly. But at the collector exit and the chimney inlet, as the flow section changes significantly, S may change significantly as well.

If the chimney is as high as 1000 m or higher, both the density inside the chimney and in the environment could not be regarded as constant. Thereby, introducing the so-called bulk air density in the system and the environment:

$$\rho_0 = \frac{\int_0^H \rho_{0,x} dx}{\int_0^H dx}, \quad \rho = \frac{\int_0^H \rho_{i,x} dx}{\int_0^H dx}$$

By using the bulk air densities in the deduction above, we can also get the similar form with Eqs. (2.9) and (2.10).

3.3 DRIVING FORCE

According to the above analysis, the minimum of S is the integration of pressure differences from the bottom to the chimney exit. As the system is connected with the environment by the collector inlet and the chimney exit, the pressure difference becomes the driving force to impel the air to flow inside the system. Thereby, according to Eq. (3.9), we get:

$$\Delta p = |S_{min}| = |S_0| = (\rho_o - \rho)gH \tag{3.11}$$

Eq. (3.11) indicates that the driving force is the absolute value of the relative static pressure at the bottom of the chimney, which can be expressed as the product of air density, gravity, and the chimney height. Unfortunately, this expression cannot give detailed information of the factors which have an effect on the driving force. As to different dimensions of SCPPSs, the collector radius, chimney radius, and solar radiation can also have significant effect on the heating, temperature difference, and air density difference. Hence, it is necessary to give a further discussion of Eq. (3.11).

Taken into account the steady flow with constant solar radiation, radiation heat transfer between the walls and the air inside the system is converted into the convection heat transfer between the soil, canopy, and the air. Evaluation of energy equilibrium along the radius of the collector and in the flow direction is carried out by dividing the collector into several concentric sections. For one of the sections, the energy equilibrium equation can be written:

$$-c_p \dot{m} \frac{dt_j}{dr} = 2\pi r h_{g,a}(T_{g,j} - T_j) + 2\pi r h_{c,a}(T_{c,j} - T_j)$$

where $h_{g,a}$ and $h_{c,a}$ are the converted convective heat transfer coefficients of the soil and the canopy to the air, respectively, and T_g and T_c are the temperature of the soil and canopy, respectively. Because the air flows to the center of the collector, air temperature decreases with the increase of radius. Thereby there is a negative sign on the left hand side of the equation. If we consider that all the solar radiation is absorbed by the air inside the system, the energy can be converted to the heat transfer from the soil to the air:

$$-c_p \dot{m} \frac{dT_j}{dr} = (1 + C_1) 2\pi r h_{g,a} (T_{g,j} - T_j) = q 2\pi r$$

Then if we integrate the equation above through the whole collector, the temperature profile along the collector can be written:

$$T - T_0 = \frac{\pi q}{c_p \dot{m}} \left(R_{coll}{}^2 - r^2 \right) \tag{3.12}$$

where R_{coll} is the collector radius. It is noted that the maximum temperature difference in the collector should be when $r = 0$ in Eq. (3.12).

The density difference may be expressed in terms of the volume coefficient of expansion, defined by:

$$\beta = \frac{\rho_0 - \rho}{\rho (T - T_0)} \tag{3.13}$$

Substituting Eqs. (3.12) and (3.13) into Eq. (3.11), we get:

$$\Delta p = \frac{\pi g \rho_0 \beta_0 q}{c_p \dot{m}} H R_{coll}{}^2 \tag{3.14}$$

Neglecting the viscosity of the flow, and substituting the Continuity Equation and Bernoulli Equation into Eq. (3.14), we can obtain:

$$\Delta p = \left(\frac{g^2}{2c_p} \right)^{\frac{1}{3}} \cdot \left(\frac{\rho_0{}^2 \beta_0{}^2}{\rho} \right)^{\frac{1}{3}} \cdot \left(\frac{H}{r_{chim}{}^2} \right)^{\frac{2}{3}} \cdot R_{coll}{}^{\frac{4}{3}} \cdot q^{\frac{2}{3}} \tag{3.15}$$

where r_{chim} is the chimney radius. From Eq. (3.15), we can see that, for the natural convection in the SCPPS, the driving force depends on the chimney height and fluid properties. However, it also depends intensively on the solar radiation and the other dimensions, such as the chimney radius and collector radius.

The driving force can be expressed by a function of the collector inlet, but for each solar chimney power plant, the collector inlet changes very little.

3.4 POWER OUTPUT AND EFFICIENCY

After successfully deriving the expression of the driving force, we can easily obtain the expressions of other solar chimney performance parameters.

Substituting Eq. (3.15) into Eq. (3.14), we can get the mass flow rate:

$$\dot{m} = \pi \left(\frac{2g\rho\rho_0 \beta_0}{c_p} \right)^{\frac{1}{3}} \cdot H^{\frac{1}{3}} \cdot r_{chim}{}^{\frac{4}{3}} \cdot R_{coll}{}^{\frac{2}{3}} \cdot q^{\frac{1}{3}} \tag{3.16}$$

The maximum power output can be expressed as the product of the driving force and the volumetric flow rate:

$$P_{max} = \frac{\Delta p \dot{m}}{\rho}$$

Substituting Eq. (3.14) to the equation above yields:

$$P_{max} = \frac{\rho_0}{\rho} \frac{\pi g}{c_p T_0} H R_{coll}^2 q \tag{3.17}$$

Eq. (3.17) shows that the air properties, the chimney height, the collector radius, and the solar radiation have significant effects on the maximum power output. Compared with the formulation advanced by Haaf [1], Eq. (3.17) gives a detailed description of factors which influence the maximum power output.

The maximum efficiency of the system can be expressed by:

$$\eta_{max} = \frac{P_{max}}{Q} = \frac{P_{max}}{\pi R_{coll}^2 q}$$

Substituting Eq. (3.17) to the equation above, we get:

$$\eta_{max} = \frac{\rho_0}{\rho} \frac{g H}{c_p T_0} \tag{3.18}$$

Eq. (3.18) demonstrates the functional dependence of the system maximum efficiency on the air density, the inlet temperature and the chimney height. Eq. (3.18) gives a more accurate expression of the maximum efficiency of the system compared with the results by Schlaich [13] and Gannon and Backström [10].

3.5 RESULTS AND DISCUSSIONS

To validate the theoretical analysis, the relative static pressure distributions are compared with the numerical simulation results of the prototype in Manzanares, Spain. Later, solar chimney models with different geometrical parameters were simulated to obtain the flow and temperature contours. During the simulation, the conservation equations for mass, momentum and energy, and standard $\kappa - \varepsilon$ turbulent equations are selected [16], and natural convection and gravity effect are taken into consideration. The main boundary conditions are shown in Table 3.1 for energy conservation equations, and the temperature profiles of the ground and the canopy in Fig. 3.2 are different parabolic functions of the radius, and the functions will vary with different solar radiations.

Figs. 3.3−3.5 show, when the solar radiation is 800 W/m², the results of 3D numerical simulations, which are the axis section flow and temperature distributions of the Spanish prototype. From Fig. 3.3, we can find that the relative static pressure distributions are negative in the system, which show good agreement

Table 3.1 Boundary Conditions and Model Parameters

Place	Type	Value
Surface of the ground	Wall	$T = f(r)$K
Surface of the canopy	Wall	$T = g(r)$K
Surface of the chimney	Wall	$q_{chim} = 0$ W/m^2
Inlet of the collector	Pressure inlet	$S = 0$ Pa, $T_0 = 293.15$K
Outlet of the chimney	Pressure outlet	$S = 0$ Pa

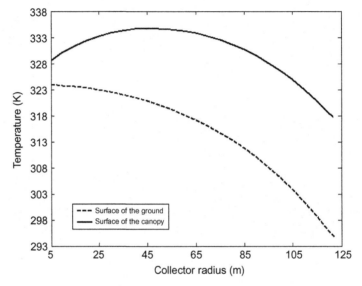

FIGURE 3.2

Temperature profiles of the ground and canopy.

with Eq. (3.10). But the minimum of relative static pressure does not lie at the location of 0 m marked A in Fig. 3.3, on the contrary, there is a large value at this location. The reasons are as follows: The air flowing in all directions causes a large swirl at this location which may result in a much smaller velocity and large static pressure. The real chimney bottom lies in the location labeled B in Fig. 3.3. At this location, the flow section changes greatly which causes the minimum relative static pressure.

The relative static pressure changes greatly at the bottom of the chimney, which shows that the largest pressure gradients also lie in this location. If an axis-based turbine is set up at this location, the static pressure difference can be used to drive the turbine and ultimately can be converted into electric energy.

From Figs. 3.6 and 3.7, we can see that the calculations done on the numerical simulation are consistent with the technical data of the Spanish prototype [1].

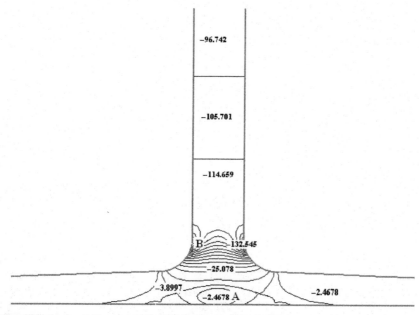

FIGURE 3.3

Relative static pressure distribution in the Spanish prototype (Pa).

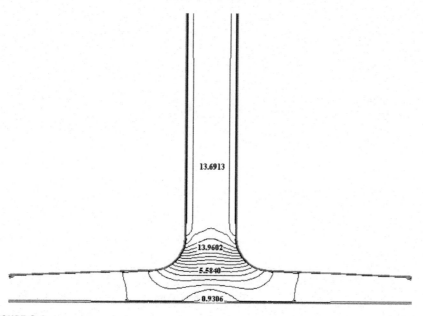

FIGURE 3.4

Velocity distribution in the Spanish prototype (m/s).

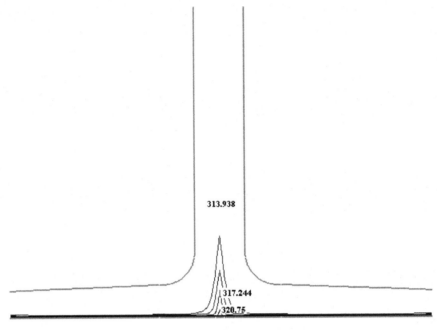

FIGURE 3.5

Temperature distribution in the Spanish prototype (K).

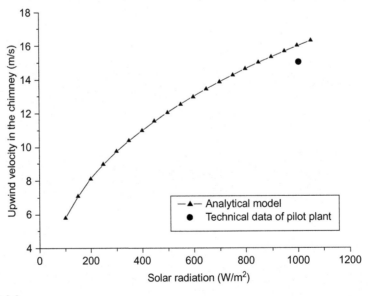

FIGURE 3.6

Comparison of velocity between analytical model and technical data.

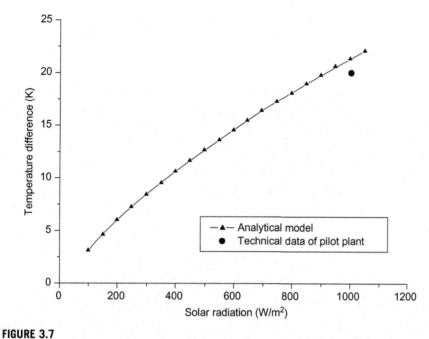

FIGURE 3.7

Comparison of temperature difference between analytical model and technical data.

When the solar radiation is 1000 W/m², the upwind velocity inside the chimney reaches about 16 m/s, and the temperature difference of the system with no-load condition simulation surpasses 20°C. The results are a little larger than the technical data only because it is supposed in this analytical model that the solar radiation is fully absorbed by the air inside the system.

From Figs. 3.3–3.5, the large swirl which lies in the collector outlet causes very low velocity, high pressure, and high temperature, which will cause large energy losses of the system. Thereby, a modification of the prototype model according to the temperature distribution shown in Fig. 3.5 was done to achieve a much more uniform temperature distribution. Figs. 3.8–3.10 show the flow and temperature contours of the new model. As a result, the local relative static pressure and velocity change significantly at the bottom of the chimney, and the temperature distribution becomes more uniform, but the main velocity, temperature, and pressure in the chimney have no appreciable changes. The larger pressure difference shown in Fig. 3.8 will be positive to the energy conversion process (from thermal and potential energy to kinetic energy) by the turbine. The results suggest that it is feasible to increase the power output by modifying the local configuration of the solar chimney power plant.

Fig. 3.11 shows the relative static pressure profiles at $x = 1$ m in the collector, which indicates that the relative static pressure is negative and decreases along

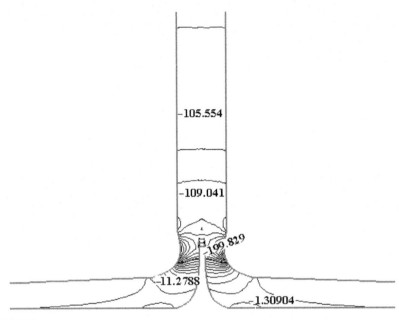

FIGURE 3.8

Relative static pressure distribution in the betterment model (Pa).

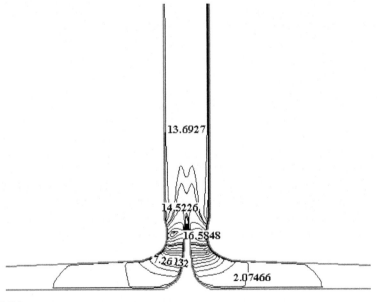

FIGURE 3.9

Velocity distribution in the betterment model (m/s).

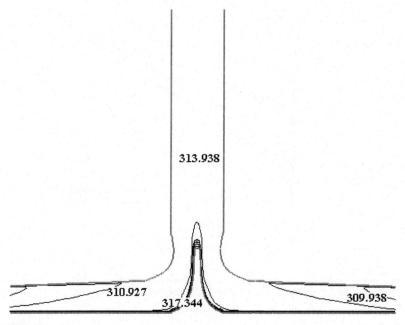

FIGURE 3.10

Temperature distribution in the betterment model (K).

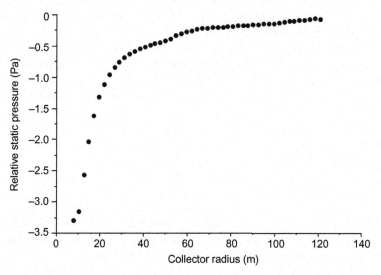

FIGURE 3.11

Relative static pressure profile in the collector at $x = 1$ m.

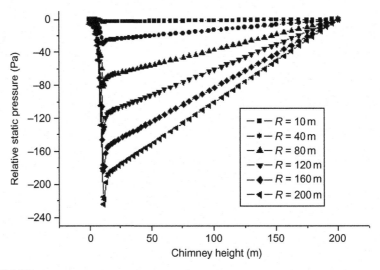

FIGURE 3.12

The effect of collector radius on the relative static pressure.

the flow direction. From this figure, we can see that static pressure is less than that of the environment at the same altitude.

Fig. 3.12 shows the relative static pressure profiles with different collector radii but the same chimney height, in which the profile marked $R = 120$ is the prototype of the Spanish model. From this figure, the minimum relative static pressure decreases with the increase of the collector radius. In addition, the relative static pressure profiles in the $0 < H < 10$ m range lie in the collector, the minimum values lie at the bottom of the chimney. The relative static pressure increases along the chimney height, which shows great agreement with Eq. (3.10).

Fig. 3.13 shows the relative static pressure profiles with different chimney heights but the same collector radius, in which the profile marked $H = 200$ m is the prototype of the Spanish model. From this figure, the minimum relative static pressure decreases with the increase of the chimney height, which shows great agreement with Eq. (3.9).

Figs. 3.14−3.16 show the influence of solar radiation, collector radius, chimney height, and chimney radius on the driving force of the system. Fig. 3.14 shows the driving forces in the chimney with the same chimney height but different collector radii, and Fig. 3.15 shows the driving forces with the same collector radius but different chimney heights, while Fig. 3.16 shows the driving forces with the same collector radius and chimney height but different chimney radii.

When the solar radiation is constant, with an increase of the collector radius, the area of the air collector increases, the air temperature in the system increases, which results in a decrease of the air density. Therefore, the relative static

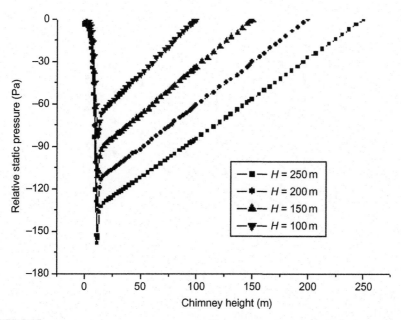

FIGURE 3.13

The effect of chimney height on the relative static pressure.

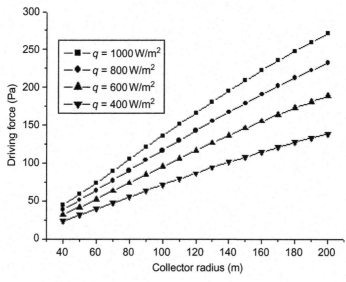

FIGURE 3.14

The effect of collector radius on the driving force.

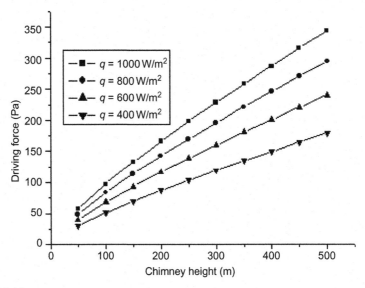

FIGURE 3.15

The effect of chimney height on the driving force.

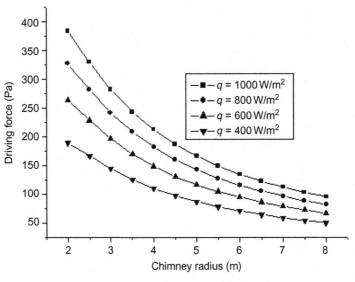

FIGURE 3.16

The effect of chimney radius on the driving force.

pressure in the system increases, and the driving force increases accordingly. Similarly, when the solar radiation increases, the air density will decrease for the same solar chimney model, and the driving force will also increase. The increase of the chimney height will cause an increase of the driving force, which has been validated by theoretical investigations. Fig. 3.15 shows reasonable agreement with Eq. (3.15).

The increase of the chimney radius will cause an increase of the mass flow rate, therefore, the air density will also increases. The increase of the air density will result in the decrease of the air density difference between the chimney and the environment. Thereby, the driving force will decrease, which is also consistent with Eq. (3.15).

Figs. 3.17−3.19 show the effect of solar radiation, collector radius, the chimney height, and the chimney radius on the power output of the system. The three figures indicate that solar radiation has a significant effect on the power output of the system.

As analyzed above, when the solar radiation is constant, with an increase of the collector radius, the area of the air collector increases with the radius by a parabolic relationship, which results in a dramatic increase of the solar energy absorbed by the air. As a result, the air has higher temperature, less density, larger driving force, and eventually a greater power output.

Similarly, an increase of the chimney height causes an increase of the mass flow rate, and the driving force as well. The driving force can be used to drive the axis-based turbine. The increase of driving force will result in the increase of

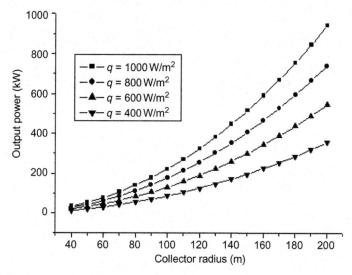

FIGURE 3.17

The effect of collector radius on the power output.

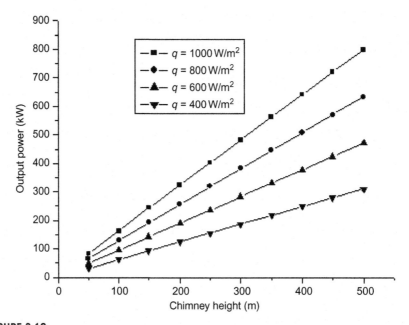

FIGURE 3.18

The effect of chimney height on the power output.

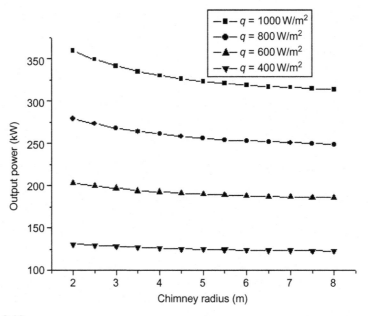

FIGURE 3.19

The effect of chimney radius on the power output.

the pressure difference across the turbine. Therefore, the chimney height has a direct effect on the power output of the system. Figs. 3.17 and 3.18 show reasonable agreement with Eq. (3.17).

The effect of chimney radius on the power output is complicated: the increase of chimney radius will cause the increase of mass flow rate, but also the increase of air density and the decrease of the driving force. This latter effect overshadows the effect of the mass flux increase. Therefore, the power output decreases slightly with the increase of the chimney radius.

Figs. 3.20–3.22 show the effect of collector radius, the chimney height, and chimney radius on the system efficiency. The curves of the former efficiency in these figures were obtained according to the formulation advanced by Schlaich [13] and Gannon and Backström [10].

The former efficiency advanced by Schlaich [13] and Gannon and Backström [10] can be influenced only by the chimney height as shown, but it is shown in the following three figures that the solar radiation, the collector radius, and the chimney radius can also have an effect on the maximum efficiency. The maximum efficiency will increase with the increase of solar radiation and collector radius, but decrease with the increase of chimney radius.

It is easy to see from Fig. 3.21 that the chimney height has a significant effect on the maximum efficiency. When the chimney height reaches 300 m, the maximum efficiency will reach 1%. It will surpass 3% when the chimney height reaches 1000 m, which can be used as a large-scale SCPPS.

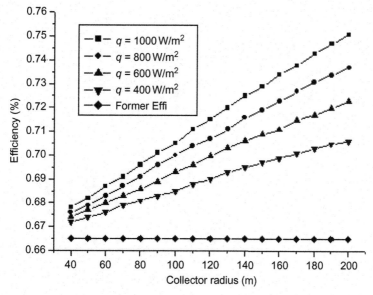

FIGURE 3.20

The effect of collector radius on the efficiency.

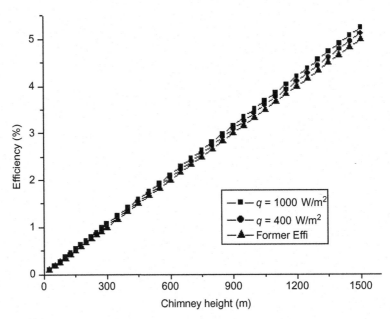

FIGURE 3.21

The effect of chimney height on the efficiency.

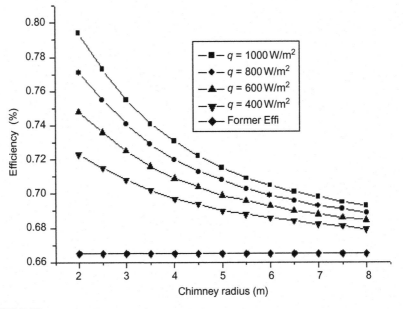

FIGURE 3.22

The effect of chimney radius on the efficiency.

In addition, the numerical simulation results also show the influence of chimney radius on the relative static pressure, driving force, power output, and the system efficiency. As shown in Fig. 3.22, when the solar radiation is constant, the effect of chimney radius on the performance of the solar chimney power plant is complicated: the increase of chimney radius will cause the increase of mass flow rate, but the increase of air density and the decrease of the driving force as well. This latter effect overshadows the effect of mass flux increase. Therefore, the power output and efficiency decrease slightly with the increase of the chimney radius.

3.6 CONCLUSIONS

The objective of this chapter was to evaluate the static pressure distribution and driving force of the solar chimney power plant theoretically. The relative static pressure was developed to analyze the correlation of static pressure between the system and the environment to estimate the pressure difference and the driving force of the system. Later, a detailed model was developed to estimate the influence of driving force by the dimensions of the solar chimney power plant and solar radiation. Then, the theoretical analysis was validated with numerical simulations of different model parameters. The theoretical analysis and numerical simulation results show that:

1. The relative static pressure in the whole system is negative, and the minimum value lies at the bottom of the chimney. The pressure difference may become larger by shrinking the flow section at the conjunction of the collector and chimney.
2. Driving force is the absolute value of the minimum value of relative static pressure. The solar radiation and the geometry of the solar chimney, such as the collector radius, the chimney height, and chimney radius, can have significant effects on the driving force of the system. The driving force of the system increases with the increase of the solar radiation, the collector radius, and the chimney height, while it decreases with the increase of the chimney radius.
3. The maximum power output is the function of the solar radiation, the collector radius, chimney height, and chimney radius. Similarly, the maximum power output of the system increases with the increase of the solar radiation, the collector radius, and the chimney height, but it decreases slightly with the increase of the chimney radius.
4. The maximum efficiency is also the function of air density and chimney height. The air density is the function of the solar radiation, the collector radius, chimney height, and chimney radius. Therefore, the maximum efficiency is influenced by the solar radiation, the collector radius, chimney height, and chimney radius. But the main factor to influence the maximum

efficiency is the chimney height. When the chimney height reaches 300 m, the maximum efficiency will reach 1%. It will surpass 3% when the chimney height reaches 1000 m which can be used as a large scale SCPPS.

NOMENCLATURE

S	relative static pressure (Pa)
x	coordinate in axial direction (m)
ρ	air density (kg/m^3)
g	gravity 9.81 (m/s^2)
H	chimney height (m)
Δp	driving force (Pa)
$\dot{m}$	mass flow rate (kg·s)
A	area (m^2)
v	air velocity (m/s)
c_p	specific heat capacity of the air 1005 (J/(kg·K))
T	temperature (K)
h	heat transfer coefficient (W/(m·K))
q	solar radiation (W/m^2)
q_{chim}	heat flux through the chimney wall (W/m^2)
R_{coll}	collector radius (m)
T_0	temperature of the environment (K)
β	the volume coefficient of expansion (1/K)
r_{chim}	the chimney radius (m)

REFERENCES

[1] Haaf W, Friedrich K, Mayer G, Schlaich J. Solar chimneys. Int J Solar Energy 1983;2:3−20.

[2] Haaf W, Friedrich K, Mayer G, Schlaich J. Solar chimneys. Int J Solar Energy 1984;2:141−61.

[3] Krisst R. Energy transfer system. Altern Sources Energy 1983;63:8−11.

[4] H. Kulunk, A prototype solar convection chimney operated under izmit condition. In: Veziroglu TN, editor. Prod. 7th MICAS, Hemisphere Publishing Corporation, Miami Beach, FL, USA 1985, p. 162.

[5] Pasumarthi N, Sherif SA. Experimental and theoretical performance of a demonstration solar chimney model—part II: experimental and theoretical results and economic analysis. Int J Energ Res 1998;22:443−61.

[6] Pasumarthi N, Sherif SA. Experimental and theoretical performance of a demonstration solar chimney model—part I: mathematical model development. Int J Energ Res 1998;22:277−88.

[7] Padki MM, Sherif SA. On a simple analytical model for solar chimneys. Int J Energ Res 1999;23:345−9.

[8] Lodhi MAK. Application of helio-aero-gravity concept in producing energy and suppressing pollution. Energ Convers Manage 1999;40:407−21.

[9] Bernardes MAD, Valle RM, Cortez MFB. Numerical analysis of natural laminar convection in a radial solar heater. Int J Therm Sci 1999;38:42—50.

[10] Gannon AJ, von Backström TW. Solar chimney cycle analysis with system loss and solar collector performance. J Sol Energ-T Asme 2000;122:133—7.

[11] Bernardes MAD, Voss A, Weinrebe G. Thermal and technical analyses of solar chimneys. Sol Energy 2003;75:511—24.

[12] Pastohr H, Kornadt O, Gurlebeck K. Numerical and analytical calculations of the temperature and flow field in the upwind power plant. Int J Energ Res 2004;28:495—510.

[13] Schlaich J, Bergermann R, Schiel W, Weinrebe G. Design of commercial solar updraft tower systems—utilization of solar induced convective flows for power generation. J Sol Energ-T Asme 2005;127:117—24.

[14] Ming TZ, Liu W, Xu GL, Yang K. Thermodynamic analysis of solar chimney power plant system. J Huazhong Univ Sci Technolog 2005;33:1—4.

[15] Liu W, Ming TZ, Yang K, Pan Y. Simulation of characteristic of heat transfer and flow for MW-graded solar chimney power plant system. J Huazhong Univ Sci Technolog 2005;33:5—7.

[16] Tao WQ. Numerical heat transfer. Xi'an, China: Xi'an Jiaotong University Press; 2001.

Fluid flow and heat transfer of solar chimney power plant*

4

Tingzhen Ming[1,2], Guoliang Xu[2], Yuan Pan[3], Fanlong Meng[2] and Cheng Zhou[2]

[1]School of Civil Engineering and Architecture, Wuhan University of Technology, Wuhan, P.R. China [2]School of Energy and Power Engineering, Huazhong University of Science and Technology, Wuhan, P.R. China [3]School of Electrical and Electric Engineering, Huazhong University of Science and Technology, Wuhan, P.R. China

CHAPTER OUTLINE

*Part of this chapter was published in Energy Conversion and Management.

4.1 INTRODUCTION

Because of the pressure of reducing carbon dioxide emissions, it is becoming more and more urgent for China to carry out various applications and research on new energy generating technologies. The solar chimney power plant system (SC), which has the following advantages while compared with the traditional power generation systems: easier to design, more convenient to draw materials, higher operational reliability, fewer running components, more convenient maintenance and overhaul, lower maintenance expense, no environmental contamination, continuous stable running, longer operational lifespan, was first proposed in the late 1970s by Professor Jörg Schlaich and tested with a prototype model in Manzanares, Spain, in the early 1980s [1,2]. It has the potential to meet the power needs of developing countries and territories, especially in deserts where there is a shortage of electric power, and has extensive application prospects.

Since the SC systems could make significant contributions to the energy supply of those countries where there is plenty of unutilized desert land, in recent years, many researchers have made research reports on this technology and have carried out tracking studies on SC systems. Haaf et al. [1,2] provided fundamental studies for the Spanish prototype in which the energy balance, design criteria, and cost analysis were discussed, and a report on the preliminary test results of the SC system was made. Some small-scale solar chimney devices with power outputs of no more than 10 W were also reported but these could only validate the feasibility and principle of the solar chimney system [3,4].

Different mathematical models based on 1-D thermal equilibrium were advanced by several researchers, Pasumarthi and Sherif [5] developed a mathematical model to study the effects of various environment conditions and geometry on the flow and heat transfer characteristics and output power of the solar chimney. Bernardes et al. [6] established a rounded mathematic model including the collector, chimney, and turbine of an SC system on the basis of the energy-balance principle. Bilgen and Rheault [7] designed a SC system for power production at high latitudes and evaluated its performance. Pretorius and Kröger [8] evaluated the influence of a developed convective heat transfer equation, more accurate turbine inlet loss coefficient, quality collector roof glass, and various types of soil on the performance of a large scale SC system.

Gannon and Von Backström [9] presented an air standard cycle analysis of the SC for the calculation of limiting performance, efficiency, and relationship between main variables including chimney friction, system, turbine, and exit kinetic energy losses. Gannon and Von Backström [10] presented an experimental investigation of the performance of a solar chimney turbine. The measured results showed that the solar chimney turbine presented has a total-to-total efficiency of 85−90% and total-to-static efficiency of 77−80% over the design range. Later, the same authors [11] presented analytical equations in terms of turbine flow and load coefficient and degree of reaction, to express the influence of each coefficient on turbine efficiency. Bernardes et al. [12] developed a thermal and technical analysis to estimate the power output and examine the effect of various ambient conditions and structural dimensions on the power output.

Pastohr et al. [13] carried out a two-dimensional steady-state numerical simulation study on the whole SC system, which consists of the energy storage layer, the collector, the turbine, and the chimney, and obtained the distributions of velocity, pressure, and temperature inside the collector. Ming et al. [14] developed a comprehensive model to evaluate the performance of a SC system, in which the effects of various parameters on the relative static pressure, driving force, power output, and efficiency have been further investigated. Ming et al. [15,16] established different mathematical models for the collector, the chimney, and the energy storage layer and analyzed the effect of solar radiation on the heat storage characteristic of the energy storage layer. Ming et al. [17] carried out numerical simulations on the SC systems coupled with a 3-blade turbine using the Spanish prototype as a practical example and presented the design and simulation of a MW-graded SC system with a 5-blade turbine, the results of which show that the coupling of the turbine increases the maximum power output of the system and the turbine efficiency is also relatively rather high.

The previous research reports indicate that, provided the considerable scale of an SC system, carrying out numerical simulations for it before setting up the commercial system would be an effective method to make predictions of the characteristic parameters and operating characteristics of the system. Apparently, mathematical models based on 1-D thermal equilibrium cannot give detailed descriptions on the temperature, velocity, and pressure distributions of the whole system. During the two-dimension simulation for an SC system carried out by Pastohr et al. [13], the pressure drop across the turbine was preset according to the Beetz power limit theory, as the turbine of the SC system is pressure-based which is similar to the turbine of a hydraulic power station but different from that of a traditional wind power station, therefore, different pressure drops across the turbine can be preset to analyze the power output characteristic of the SC system. In addition, the existing numerical simulation results did not analyze the influences of solar radiation, turbine pressure drop and efficiency on the flow and heat transfer, output power and energy loss of the SC system. Based on this, two-dimension numerical simulations for the Spanish solar chimney power plant prototype, containing an energy storage layer, a collector, a turbine, and the chimney, will be carried out in this paper to analyze the influence of pressure drop across the turbine and solar radiation on items such as flow, heat transfer, energy loss, and power output characteristics of the SC system.

4.2 THEORETICAL MODELS

4.2.1 PHYSICS MODEL

An SC system with the same fundamental dimensions as the Spanish SC prototype will be built in Wuhan, China. Taking the fundamental dimensions of the

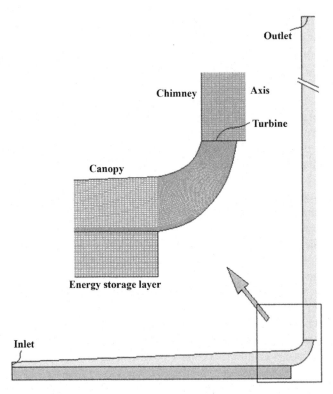

FIGURE 4.1

Physical model of the SC prototype.

Spanish SC system shown in Fig. 4.1 as a practical example [2], geometric parameters of the prototype are shown as follows: height of chimney, 200 m; diameter of chimney, 10 m; radius of collector, 122 m; height from the inlet to its center, 2−6 m; thickness of the energy storage layer, 5 m. The collector and the chimney are smoothly connected in order to reduce the energy loss caused by resistance, and in addition a shrink flow passage is designed at the bottom of the collector. The material used for the collector canopy is transparent glass, while the energy storage layer uses the soil that can be regarded as porous media.

4.2.2 MATHEMATICAL MODEL

In natural convection, the strength of the buoyancy-induced flow is measured by the Rayleigh number, defined as follows:

$$Ra = \frac{g\beta\Delta TL^3}{a\nu} \tag{4.1}$$

where, ΔT is the maximum temperature difference of the system. L, a, and β are the mean collector height, the thermal diffusivity, and the thermal expansion coefficient, respectively. The whole collector and chimney analysis shows that $Ra > 10^{10}$, therefore, fluid flow in the regions may be turbulent. Accordingly, the axisymmetric mathematical model including the continuity equation, Navier-Stocks equation, energy equation, and $\kappa - \varepsilon$ equations used to describe the problem are as follows:

$$\frac{\partial(\rho u)}{\partial x} + \frac{1}{r}\frac{\partial(r\rho v)}{\partial r} = 0 \tag{4.2}$$

$$\frac{\partial}{\partial x}(\rho uu) + \frac{1}{r}\frac{\partial}{\partial r}(r\rho vu) = \frac{\partial}{\partial x}\left((\mu + \mu_t)\frac{\partial u}{\partial x}\right) + \frac{1}{r}\frac{\partial}{\partial r}\left((\mu + \mu_t)r\frac{\partial u}{\partial r}\right)$$
$$+ \rho g\beta(T - T_0) + \frac{\partial}{\partial x}\left((\mu + \mu_t)\frac{\partial u}{\partial x}\right) + \frac{1}{r}\frac{\partial}{\partial r}\left((\mu + \mu_t)r\frac{\partial v}{\partial x}\right) \tag{4.3}$$

$$\frac{\partial}{\partial x}(\rho uv) + \frac{1}{r}\frac{\partial}{\partial r}(r\rho vv) = \frac{\partial}{\partial x}\left((\mu + \mu_t)\frac{\partial v}{\partial x}\right) + \frac{1}{r}\frac{\partial}{\partial r}\left((\mu + \mu_t)r\frac{\partial v}{\partial r}\right)$$
$$+ -\frac{\partial p}{\partial r} + \frac{\partial}{\partial x}\left((\mu + \mu_t)\frac{\partial u}{\partial r}\right) + \frac{1}{r}\frac{\partial}{\partial r}\left((\mu + \mu_t)r\frac{\partial v}{\partial r}\right) - \frac{2(\mu + \mu_t)v}{r^2} \tag{4.4}$$

$$\frac{\partial(uT)}{\partial x} + \frac{1}{r}\frac{\partial(rvT)}{\partial r} = \frac{1}{\rho}\frac{\partial}{\partial x}\left(\left(\frac{\mu}{Pr} + \frac{\mu_t}{\sigma_T}\right)\frac{\partial T}{\partial x}\right) + \frac{1}{\rho r}\frac{\partial}{\partial r}\left(\left(\frac{\mu}{Pr} + \frac{\mu_t}{\sigma_T}\right)r\frac{\partial T}{\partial r}\right) \tag{4.5}$$

$$\frac{\partial(u\kappa)}{\partial x} + \frac{1}{r}\frac{\partial(rv\kappa)}{\partial r} = \frac{1}{\rho}\frac{\partial}{\partial x}\left(\left(\mu + \frac{\mu_t}{\sigma_\kappa}\right)\frac{\partial\kappa}{\partial x}\right) + \frac{1}{\rho r}\frac{\partial}{\partial r}\left(\left(\mu + \frac{\mu_t}{\sigma_\kappa}\right)r\frac{\partial\kappa}{\partial r}\right) + G_\kappa - \varepsilon \tag{4.6}$$

$$\frac{\partial(u\varepsilon)}{\partial x} + \frac{1}{r}\frac{\partial(rv\varepsilon)}{\partial r} = \frac{1}{\rho}\frac{\partial}{\partial x}\left(\left(\mu + \frac{\mu_t}{\sigma_\varepsilon}\right)\frac{\partial\varepsilon}{\partial x}\right) + \frac{1}{\rho r}\frac{\partial}{\partial r}\left(\left(\mu + \frac{\mu_t}{\sigma_\varepsilon}\right)r\frac{\partial\varepsilon}{\partial r}\right) + \frac{\varepsilon}{\kappa}(C_1 G_\kappa - C_2\varepsilon)$$
$$\tag{4.7}$$

where, G_κ represents the generation of turbulence kinetic energy due to the mean velocity gradients defined as: $G_\kappa = -\mu_t\left(2\left(\left(\frac{\partial u}{\partial x}\right)^2 + \left(\frac{\partial v}{\partial r}\right)^2 + \left(\frac{v}{r}\right)^2\right) + \left(\frac{\partial u}{\partial r} + \frac{\partial v}{\partial x}\right)^2\right)$. σ_T, σ_κ, and σ_ε are the turbulent Prandtl numbers for T, κ, and ε, respectively, and c_1 and c_2 are two constants for the turbulent model: $c_1 = 1.44$, $c_2 = 1.92$, $\sigma_T = 0.9$, $\sigma_\kappa = 1.0$, $\sigma_\varepsilon = 1.3$. $\mu_t = \frac{c_\mu \rho \kappa^2}{\varepsilon}$, and $c_\mu = 0.09$.

The heat transfer and flow in the energy storage layer may be very complicated, and it is necessary to take into account the collector, the chimney, and the storage medium as a whole system. As the material used for the energy storage layer can be regarded as a porous medium, the Brinkman−Forchheimer Extended Darcy model [18] is used to describe the flow in the convective porous-layer, which can be expressed as follows:

$$\frac{\partial u}{\partial x} + \frac{1}{r}\frac{\partial(rv)}{\partial r} = 0 \tag{4.8}$$

$$\frac{\rho}{\varphi}\frac{\partial u}{\partial t} + \frac{\rho}{\varphi^2}\left(\frac{\partial(uu)}{\partial x} + \frac{1}{r}\frac{\partial(rvu)}{\partial r}\right) = -\frac{\partial p}{\partial x} + \frac{\partial}{\partial x}\left(\mu_m\frac{\partial u}{\partial x}\right)$$

$$+ \frac{1}{r}\frac{\partial}{\partial r}\left(r\mu_m\frac{\partial u}{\partial r}\right) - \frac{\mu u}{K} - \frac{\rho F}{\sqrt{K}}\sqrt{u^2+v^2}u + \rho g\beta(T-T_e) \tag{4.9}$$

$$\frac{\rho}{\varphi}\frac{\partial v}{\partial t} + \frac{\rho}{\varphi^2}\left(\frac{\partial(uv)}{\partial z} + \frac{1}{r}\frac{\partial(rvv)}{\partial r}\right) = \frac{\partial}{\partial z}\left(\mu_m\frac{\partial v}{\partial z}\right) +$$

$$\frac{1}{r}\frac{\partial}{\partial r}\left(r\mu_m\frac{\partial v}{\partial r}\right) - \mu_m\frac{v}{r^2} - \frac{\mu u}{K} - \frac{\rho F}{\sqrt{K}}\sqrt{u^2+v^2}v \tag{4.10}$$

$$\rho_m c_{p,m}\left(\frac{\partial T}{\partial t} + \frac{\partial(uT)}{\partial z} + \frac{1}{r}\frac{\partial(rvT)}{\partial r}\right) = \frac{\partial}{\partial z}\left(\lambda_m\frac{\partial T}{\partial z}\right) + \frac{1}{r}\frac{\partial}{\partial r}\left(r\lambda_m\frac{\partial T}{\partial r}\right) \tag{4.11}$$

where, φ, ρ_m, $c_{p,m}$, μ_m, and λ_m are the porosity, apparent density, specific capacity, dynamic viscosity, and apparent thermal conductivity of the porous medium, respectively: $\rho_m = (1-\varphi)\rho_s + \varphi\rho_a$, $c_{p,m} = (1-\varphi)c_{p,s} + \varphi c_{p,a}$, $\lambda_m = (1-\varphi)\lambda_s + \varphi\lambda_a$, $\mu_m = \mu/\varphi$; the parameters with subscripts s and a denote the corresponding parameters of the solid and air in the energy storage layer, respectively. K, F, and d_b are the permeability, the inertia coefficient, and the particle diameter of the energy storage layer, respectively.

$$K = d_b^2\varphi^3/(175(1-\varphi)^2) \tag{4.12}$$

$$F = 1.75\varphi^{-1.5}/\sqrt{175} \tag{4.13}$$

4.2.3 BOUNDARY CONDITIONS AND SOLUTION METHOD

Boundary conditions are set as follows: For the roof of the collector, we take the convection boundary into account, and the coefficient of convection is set as 10 W/(m² K) which can be accepted when the environment air velocity is not very large, that is, 1~2 m/s. The temperature of the environment is set as 293K; the inlet of the collector is set as the pressure-inlet boundary; the chimney wall can be set as an adiabatic boundary; the chimney outlet is set as the pressure-outlet boundary; the bottom of the energy storage layer is set as the temperature-constant boundary, whose temperature is 300K. Solar radiation which projects through the transparent ceiling into the ground can be considered as a heat source for the ground thin layer [13]. In recent years, the highest solar radiation in Wuhan area is about 650 W, therefore, we can simulate the running conditions of a SC system in Wuhan on the whole by setting solar radiation as 200, 400, 600, and 800 W/m², respectively.

In addition, it is necessary to explain the reason why we consider the collector inlet and chimney outlet both as pressure boundaries and have their pressures set as 0 Pa. This is because we simultaneously take the inner and outer pressure distributions of the system into account [13]; $p_{r,i} = 0$ means that for both the inside and outside of the collector inlet, the static pressures at the same height are the same.

However, the setting of the pressure drop across the turbine in this paper differs from the processing method applied by Pastohr [13]. The turbine of an SC system, as explained earlier, belongs to a pressure-based wind turbine, the fore-and-aft air velocities are almost the same but the pressure changes significantly, and its power output does not follow the Beetz power limit theory. Thereby the output power through the turbine can be calculated according to Eq. (4.14) by presetting the pressure drop across the turbine:

$$W_t = \eta_t \cdot \Delta p \cdot V \tag{4.14}$$

where, W_t represents the shaft power output through the turbine, η_t represents the energy conversion efficiency of the turbine, which can be preset as 80%, less than the optimized data [10,11], Δp represents the pressure drop across the turbine, V represents the air volume flow rate of the system flowing through the chimney outlet. The boundary conditions for different locations are shown in Table 4.1.

The standard $k - \varepsilon$ model is applied during the numerical simulation of air flow in the collector and chimney, The material for the energy storage layer is selected as soil, and the properties of the soil are as follows: $\rho_{soil} = 1700 \text{ kg/m}^3$, $c_{p,soil} = 2016 \text{ J/(kg·K)}$, $\lambda_{soil} = 0.78 \text{ W/(m·K)}$. The porosity of the energy storage layer is selected as 0.1, as in Wuhan, China, the soil is almost close-grained. The standard wall function method is applied, the SIMPLE algorithm is applied for pressure-velocity coupling, and the momentum equation, energy equation, and other equations all apply the second-order upwind discretization scheme. The mesh number of the SC system is nearly 500,000, where we can get a grid-independent simulation result. In addition, the feasibility of this numerical method can be found in Ref. [15].

Table 4.1 Boundary Conditions

Location	Type	Value
0.1 mm top layer of the ground	Heat source	$2 \times 10^6 \sim 8 \times 10^6 \text{ W/m}^3$
Bottom of the ground	Temperature	300K
Surface of the canopy	Wall	$T_e = 293\text{K}, h = 10 \text{ W/(m}^2\text{·K)}$
Surface of the chimney	Wall	$q_{chim} = 0 \text{ W/m}^2$
Collector inlet	Pressure inlet	$p_{r,i} = 0 \text{ Pa}, T_0 = 293\text{K}$
Chimney outlet	Pressure outlet	$p_{r,o} = 0 \text{ Pa}$
Pressure drop across the turbine	Reverse fan	$0 \sim 480 \text{ Pa}, 40 \text{ Pa in interval}$

4.3 RESULTS AND DISCUSSION

Currently, solar power plant and wind power plant are the main power generation systems using renewable energy, namely solar power and wind power respectively. Turbines applied in an SC system and a wind power plant are pressure-based impulse turbines and velocity-based impulse turbines, respectively. For the former system, where the air flow is subjected to the area of the chimney, fore-and-aft velocity changes little, accompanied with a pressure drop. In the latter system, fore-and-aft pressure changes little while the velocity decreases significantly.

Figs. 4.2−4.4 show the simulation results on flow and heat transfer characteristics of the SC system when solar radiation and pressure drop across the turbine are 400 W/m^2 and 0 Pa (when turbine is not in service), respectively. As shown in Fig. 4.2, relative static pressure (defined as the difference between the inner-system static pressure and the environment static pressure at the same height) inside the system has negative values everywhere, this indicates that pressure inside the system is always lower than that of the environment. Among this, the pressure difference between the chimney bottom and the environment comes to the maximum. Meanwhile, the velocity also reaches its maximum which is over 12 m/s. On the other hand, temperature inside the system increases up to 307K, experiencing a 15K change compared with the environment temperature. Obviously, under no-load condition, this temperature variation is not very dramatic.

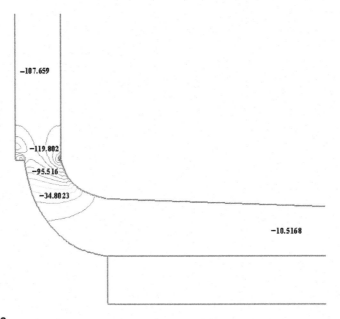

FIGURE 4.2

Relative static pressure distribution when the solar radiation and pressure drop across the turbine are 400 W/m^2 and 0 Pa, respectively (Pa).

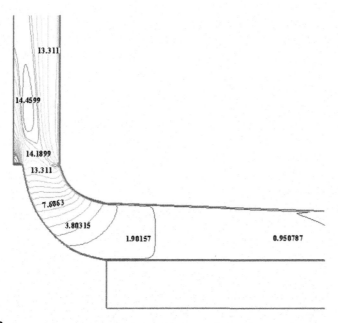

FIGURE 4.3

Velocity distribution when the solar radiation and pressure drop across the turbine are 400 W/m² and 0 Pa, respectively (m/s).

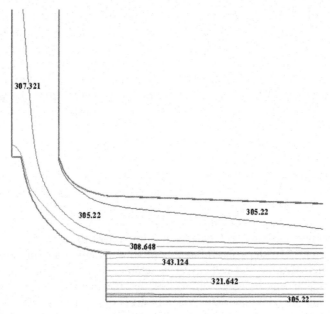

FIGURE 4.4

Temperature distribution when the solar radiation and pressure drop across the turbine are 400 W/m² and 0 Pa, respectively (K).

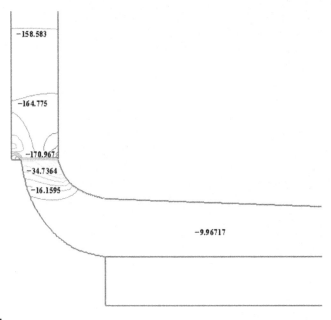

FIGURE 4.5

Relative static pressure distribution when the solar radiation and pressure drop across the turbine are 400 W/m^2 and 120 Pa, respectively (Pa).

Figs. 4.5–4.8 show the simulation results on flow and heat transfer characteristics of the SC system when solar radiation and pressure drop across the turbine are 400 W/m^2 and 120 Pa, respectively. It is found that, compared with Figs. 4.5–4.6 and 4.2, the pressure distribution inside the system has experienced dramatic variations, among which the pressure drop within the turbine area is extremely dramatic. This is because the turbine is preset a 120 Pa pressure drop during this numerical simulation, and the results as shown in Figs. 4.5 and 4.6 exactly reflect this effect.

Through comparing the two simulation results corresponding to the turbine pressure drops being 0 and 120 Pa respectively, we found that when the turbine pressure drop increases, the inside-outside pressure difference of the system increases, velocity of the air flow decreases, while the temperature of the air flow increases, for which the main reason is that the turbine pressure drop has an inverse effect on the air velocity. Thus the heating-up period is prolonged, and the temperature at the chimney outlet increases with the pressure drop across the turbine.

Figs. 4.9 and 4.10 show the influences of solar radiation and turbine pressure drop on the temperature and velocity at the chimney outlet. As shown in the figures, when the turbine pressure drop increases, the chimney outlet temperature increases, while the chimney outlet velocity decreases gradually. This is because the increase of turbine pressure drop causes a block effect to the air flow within

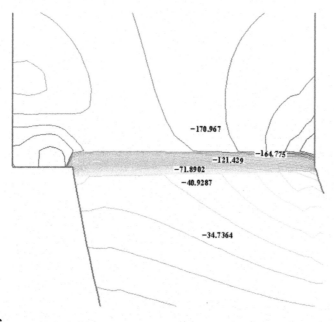

FIGURE 4.6

Local relative static pressure distribution near the turbine when the solar radiation and pressure drop across the turbine are 400 W/m^2 and 120 Pa, respectively (Pa).

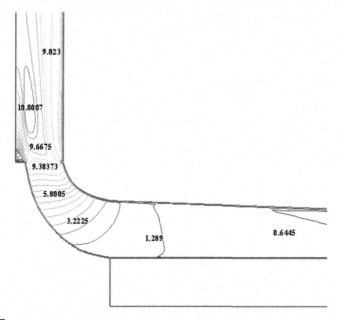

FIGURE 4.7

Velocity distribution when the solar radiation and pressure drop across the turbine are 400 W/m^2 and 120 Pa, respectively (m/s).

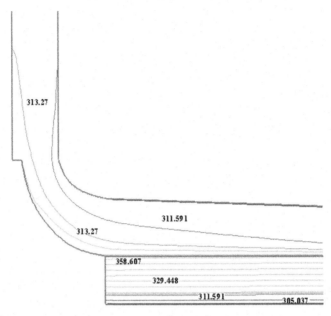

FIGURE 4.8

Temperature distribution when the solar radiation and pressure drop across the turbine are 400 W/m^2 and 120 Pa, respectively (K).

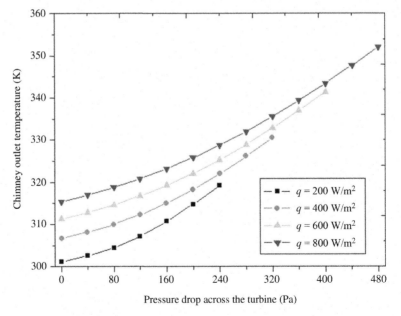

FIGURE 4.9

Influences of solar radiation and turbine pressure drop on chimney outlet temperature.

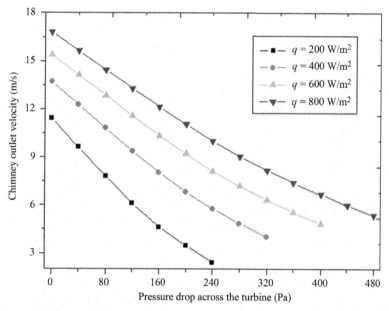

FIGURE 4.10

Influences of solar radiation and turbine pressure drop on chimney outlet velocity.

the system which makes the air flow rate of the system decrease. As a result, the heating-up time of the air inside the collector is prolonged and the chimney outlet temperature increases. Meanwhile, if the turbine pressure drop keeps constant, the air velocity increases notably with the solar radiation, which originates from the relatively high extent to which the fluid is heated during free convection process. Therefore, both the temperature and velocity of the chimney outlet increase.

In general, the air velocities before turbines are often in the scope of 8−9 m/s when wind turbines are in their operation conditions. Based on this scope, it is shown from Fig. 4.10, corresponding to solar radiation being 200, 400, 600, and 800 W/m^2, respectively, that the optimal values of the turbine pressure drop are 40, 120, 200, and 280 Pa, respectively. As shown in Fig. 4.9, the variation range for the chimney outlet temperature is 303 ∼ 333K under four optimal turbine pressure drops, and the chimney outlet temperature is 320K when the solar radiation and turbine pressure drop are 600 W/m^2 and 200 Pa, respectively.

Fig. 4.11 shows the influences of the turbine pressure drop and solar radiation on the turbine output power of the SC system, in which the efficiency of turbine is preset as 80%. As shown in Fig. 4.11, if the turbine pressure drop keeps constant, output power of the system increases with the solar radiation. This is because the increase of solar radiation will result in a notable increase of the system air volume flow rate. However, the influence of turbine pressure drop on the system output power is rather complicated. Under a small turbine pressure drop,

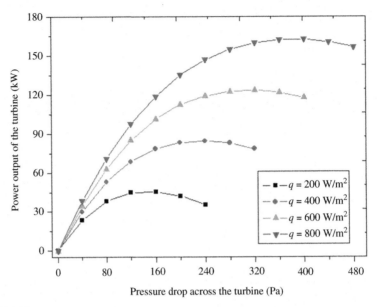

FIGURE 4.11

Influences of solar radiation and turbine pressure drop on turbine output power.

the system output power increases with the turbine pressure drop. According to Eq. (4.14), the main reason for this phenomenon is that the reduction of the air volume flow rate of the system caused by the turbine pressure drop is relatively small, making the product of the air volume flow rate and the pressure drop across the turbine present a rise trend. But when the turbine pressure drop is fairly large, the air volume flow rate decreases more significantly than the extent to which the turbine pressure drop increases, thus resulting in a reduction of the system output power.

Through comparing the simulation results in Fig. 4.11 and the experimental results of the Spanish SC prototype system [2], we found that the experimental output power of the Spanish SC prototype system is 35 kW under a solar radiation of about 750 W/m², while the maximum output power in Fig. 4.11 is higher than 40 kW under a solar radiation of only 200 W/m². The main reason for this difference is that the design of the turbine used in the Spanish SC prototype system is far from optimization. From Fig. 4.11 we can see that the output power of the system by using an optimized turbine will be much higher than that of the Spanish SC prototype system when the solar radiation is over 600 W/m². However, no system output power was reported in Ref. [13] which based its simulation for turbine operation on Beetz theory during the simulation process. On the other hand, as mentioned above, the power efficiency of a pressure-based turbine dramatically differs from that of a velocity-based free wind field turbine. The simulation results in this work take the turbine as pressure-based, and the turbine

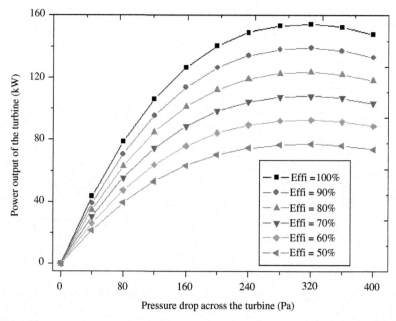

FIGURE 4.12

Influences of turbine efficiency and turbine pressure drop on output power of turbine.

efficiency is preset as 80% which is fairly easy to achieve, as shown in Refs. [10,11], as a result, this brings about relatively a dramatic difference between the theoretical simulation results and the experimental results shown in Ref. [2], which serves as a theoretical law for further optimum design of the turbine coupled with the SC system.

Fig. 4.12 shows the influences of turbine efficiency and turbine pressure drop on the system output power under a solar radiation of 600 W/m^2. As shown in this figure, besides the notable influence of turbine pressure drop on the output power of the system, the turbine efficiency also has a notable influence on the output power of the system. With the turbine pressure drop remaining constant, the higher the efficiency of turbine, the larger the output power of the system. Again through comparing the simulation results in this work with the experimental results in Ref. [2], we found that even if the turbine efficiency is only 50%, provided that that solar radiation is 600 W/m^2 and pressure drop is in the scope of 80~400 Pa, the system output power is always larger than the experimental results shown in Ref. [2]. Therefore, it is concluded that if we increase the turbine efficiency by improving the turbine structure to an optimized design, the system output power will increase remarkably. Additionally, Fig. 4.12 also indicates that the pressure-based turbine of the SC system is able to run in a relatively wide range of turbine pressure drop with a fairly large system output power.

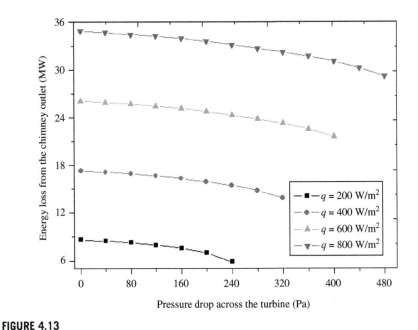

FIGURE 4.13

Energy loss from the chimney outlet caused by pressure drop across the turbine.

According to the simulation results as shown in Fig. 4.11, under a solar radiation of 600 W/m^2, the system output power is merely 120 kW, even when an ideal and optimum designed turbine with a turbine efficiency of 80% is applied. Taking the fact that the component which receives solar radiation within the system is a collector whose radius is 122 m, we can concluded that the total energy conversion efficiency from solar energy to turbine shaft output power is only 0.428%, for which the main reason is possibly excessive energy loss.

When the air flows through the chimney outlet, the total energy loss includes kinetic energy loss, gravitational potential energy, and enthalpy, among which the fractional of the kinetic energy and gravitational potential energy are far smaller than the air enthalpy flowing out of the system. These two items of macroscopic energy loss are negligible when calculating energy loss from chimney outlet. Therefore the total energy loss from the chimney outlet can be defined as the total enthalpy difference between the chimney outlet and the collector inlet. Fig. 4.13 shows the influence of turbine pressure drop and solar radiation on energy loss from the chimney outlet. As shown in this figure, energy loss from the chimney outlet is obviously rather large, it is 100–1000 times larger than the output power of the system. It indicates that the chimney outlet becomes the most important part of energy loss from the system. When the pressure drop across the turbine is small, the total energy loss from the chimney outlet accounts for 90% of the solar energy received by the system; even when the turbine pressure drop is high, the

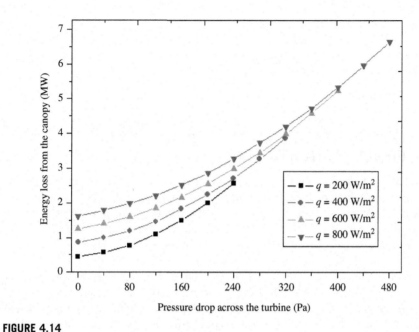

FIGURE 4.14

Energy loss from the canopy caused by pressure drop across the turbine.

energy loss from the chimney outlet still accounts for about 75% of total solar energy received by the system, thus resulting in a rather low energy conversion efficiency of the system. Furthermore, from Fig. 4.13 we can also see that energy loss from the chimney outlet increases significantly with the solar radiation but decreases with the turbine pressure drop which is mainly because the air flow rate decreases dramatically with the turbine pressure drop.

Fig. 4.14 shows the energy loss from the canopy caused by the turbine pressure drop and the solar radiation. Obviously, energy loss from the canopy increases significantly with pressure drop across the turbine. Under a solar radiation of 200 W/m^2 and a pressure drop across the turbine of 200 Pa, the total energy loss from the canopy is nearly 2 MW, namely 50 W/m^2 from the system to the environment through the canopy, which is nearly 1/4 of the solar radiation intensity; under a solar radiation intensity of 600 W/m^2 and a pressure drop across the turbine of 360 Pa, the total energy loss from the canopy is as much as 5 MW, namely 125 W/m^2, which is also nearly 1/4 of total solar radiation received. These numbers illustrate that the energy loss from the canopy is also very large, and one way to reduce energy loss from the canopy is to apply a double-layer transparent material, which can effectively reduce energy loss from the canopy.

In addition, numerical simulation results show that the portion of energy loss toward the ground through the energy storage layer is rather small, and is far smaller than the portion of energy loss through the chimney outlet and canopy.

When the solar radiation and turbine pressure drop are 600 W/m^2 and 360 Pa, respectively, the total energy loss from the bottom of the energy storage layer is only 600 kW. Therefore, the energy loss through the bottom of the energy storage layer is negligible when analyzing the total energy loss of SC system.

4.4 HELICAL HEAT-COLLECTING SOLAR CHIMNEY POWER PLANT SYSTEM

As for SC systems that already have certain scale, elements that determine the initial investment include: chimney, collector, turbine, etc. In detail, the chimney decides the system's efficiency, decreasing the chimney's height will remarkably reduce the cost of SC system but also decrease the system's efficiency. Therefore, choosing cheap materials or applying the Floating Solar chimney [19–22] is feasible in reducing the cost of the chimney. Increasing the turbine's efficiency is also another important method in reducing the cost of SC system, which is the main focus for a lot of research being undertaken [10,11,23]. Concerning the fact that the area of the collector is rather large, when the system material is selected, the cost of the collector and SC system's power increase as the area of the collector increases, considering the cost of land area and material, the area of the collector has a determinant influence on the economical efficiency of the whole system. If the area of the collector is shrunk in a proper manner while the maximum system output power remains the same under the same solar radiation, then the efficiency of the SC system can be increased, the system's economical efficiency of the system is improved, while the initial investment is reduced. In this chapter, a new type of helical heat-collecting solar chimney power generating system is proposed, and numerical analysis on the system's fluid flow and heat transfer characteristics is carried out and compared with the experimental and simulation results of the solar chimney prototype plant in Spain.

4.5 MATHEMATICAL AND PHYSICAL MODEL

4.5.1 PHYSICAL MODEL

The basic dimensions of the SC prototype plant in Spain are as follows: chimney height, 200 m; chimney radius, 10 m; collector radius, 122 m; height between the inlet of the collector and the center, 2~6 m. In order to compare the helical heat-collecting SC system with the SC prototype plant in Spain [1], the basic dimensions of the helical heat-collecting SC system established in this paper are as follows: chimney height, 200 m; chimney diameter, 10 m; collector radius, 90 m; height between the inlet of the collector and the center, 2~6 m. The planform of the system is shown as Fig. 4.15, 4 helical transparent walls, which can be made

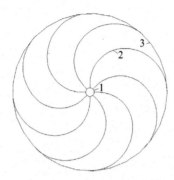

FIGURE 4.15

Planform of the collector configuration of new type of helical heat-collecting solar chimney power generating system. 1. chimney; 2. helical transparent wall; 3. collector inlet.

of the same glass material as the canopy of the collector, are set in the center—the reason for applying transparent glass as the wall material is to avoid the shading of solar radiation. Through design, the fluid is driven to flow along the helical transparent wall to the collector center, hence the actual route length along which the fluid is heated within the collector is 145 m. In order to reduce the resistance loss, the collector and the chimney are smoothly connected; moreover, shrinkage flow path is applied at the collector bottom. The canopy of the collector is made of transparent glass.

4.5.2 MATHEMATICAL MODEL

Through analysis, the fluid flow within the collector and the chimney should be a vigorous turbulence region. The relevant continuity equations, N-S equations, energy equations, and turbulence equations are as follows:

$$\frac{\partial \rho}{\partial t} + \frac{\partial(\rho u)}{\partial x} + \frac{\partial(\rho v)}{\partial y} + \frac{\partial(\rho w)}{\partial z} = 0 \tag{4.15}$$

$$\frac{\partial(\rho u)}{\partial t} + \frac{\partial(\rho uu)}{\partial x} + \frac{\partial(\rho vu)}{\partial y} + \frac{\partial(\rho wu)}{\partial z} = \rho g\beta(T - T_\infty) + \mu\left(\frac{\partial^2 u}{\partial x^2} + \frac{\partial^2 u}{\partial y^2} + \frac{\partial^2 u}{\partial z^2}\right) \tag{4.16}$$

$$\frac{\partial(\rho v)}{\partial t} + \frac{\partial(\rho uv)}{\partial x} + \frac{\partial(\rho vv)}{\partial y} + \frac{\partial(\rho wv)}{\partial z} = -\frac{\partial p}{\partial y} + \mu\left(\frac{\partial^2 v}{\partial x^2} + \frac{\partial^2 v}{\partial y^2} + \frac{\partial^2 v}{\partial z^2}\right) \tag{4.17}$$

$$\frac{\partial(\rho w)}{\partial t} + \frac{\partial(\rho uw)}{\partial x} + \frac{\partial(\rho vw)}{\partial y} + \frac{\partial(\rho ww)}{\partial z} = -\frac{\partial p}{\partial z} + \mu\left(\frac{\partial^2 w}{\partial x^2} + \frac{\partial^2 w}{\partial y^2} + \frac{\partial^2 w}{\partial z^2}\right) \tag{4.18}$$

$$\frac{\partial(\rho cT)}{\partial t} + \frac{\partial(\rho cuT)}{\partial x} + \frac{\partial(\rho cvT)}{\partial y} + \frac{\partial(\rho cwT)}{\partial z} = \lambda\left(\frac{\partial^2 T}{\partial x^2} + \frac{\partial^2 T}{\partial y^2} + \frac{\partial^2 T}{\partial z^2}\right) \tag{4.19}$$

$$\frac{\partial}{\partial t}(\rho k) + \frac{\partial}{\partial x_i}(\rho k u_i) = \frac{\partial}{\partial x_j}\left(\left(\mu + \frac{\mu_t}{\sigma_k}\right)\frac{\partial k}{\partial x_j}\right) + G_k + G_b - \rho\varepsilon + S_k \quad (4.20)$$

$$\frac{\partial}{\partial t}(\rho\varepsilon) + \frac{\partial}{\partial x_i}(\rho\varepsilon u_i) = \frac{\partial}{\partial x_j}\left(\left(\mu + \frac{\mu_t}{\sigma_\varepsilon}\right)\frac{\partial\varepsilon}{\partial x_j}\right) + C_{1\varepsilon}(G_k + C_{3\varepsilon}G_b) - C_{2\varepsilon}\rho\frac{\varepsilon^2}{k} + S_\varepsilon \quad (4.21)$$

In the equations above, G_k represents the turbulence kinetic energy generation item caused by average velocity gradient: $G_k = -\overline{\rho u_i' u_j'}\dfrac{\partial u_j}{\partial x_i}$; G_b refers to the turbulence kinetic energy generation item caused by buoyancy; σ_k and σ_ε are constants of $k - \varepsilon$ equations; β refers to the coefficient of cubic expansion: $\beta \approx 1/T$.

4.5.3 SOLVING DETERMINANT CONDITION AND SOLUTION

1. Heat balance condition of the glass surface of the collector canopy

$$Q_{g,air} + Q_{g,e} + Q_{g,stor} + Q_{g,sky} + \alpha Q_{solar} = 0 \quad (4.22)$$

In the equation above, $Q_{g,air}$ refers to the heat convection flux between the collector surface and air inside the collector. As the temperature difference between them is not big at all, hence the radiation heat transfer between them is negligible. $Q_{g,air} = A_g h_{g,air}(T_g - T_a)$, where A_g refers to the surface of the collector; $h_{g,air}$ refers to the convection heat transfer coefficient between the collector surface and the air inside the collector; T_g and T_a represent the thermodynamic temperature of the collector surface and the air inside the collector, respectively; $Q_{g,e}$ refers to the convection heat transfer flux between the collector surface and the ambient air, $Q_{g,e} = A_g h_{g,e}(T_g - T_e)$; $h_{g,e}$ refers to the convection heat transfer coefficient between the collector surface and the ambient air; T_e refers to the thermodynamic temperature of the ambient air; $Q_{g,stor}$ refers to the radiation heat transfer flux between the collector surface and the surface of the heat storage layer at the bottom of the collector, Considering the radiation heat transfer between the collector surface and the heat storage layer as radiation heat transfer between two parallel plates of the same area and ignoring the radiation loss at the collector inlet and outlet, we have: $Q_{g,stor} = A_g \sigma(T_g^4 - T_{stor}^4)$, where σ refers to Stephen Boltzmann constant; T_{stor} refers to the thermodynamic temperature of the heat storage layer; $Q_{g,sky}$ refers to the radiation heat transfer flux between the collector surface and the sky, $Q_{g,sky} = A_g \sigma(T_g^4 - T_{sky}^4)$; Q_{solar} refers to the solar radiation received by the collector; α refers to the absorptivity of the transparent material of the collector.

2. Surface condition of heat storage layer

$$Q_{stor,air} + Q_{stor,g} + Q_{stor,down} + \eta\tau Q_{solar} = 0 \quad (4.23)$$

$Q_{stor,air}$ refers to the heat transfer flux between the heat storage layer and the air inside the collector, $Q_{stor,air} = A_{stor} h_{stor,air}(T_{stor} - T_{air})$; A_{stor} refers to the heat transfer area of the heat storage layer: $A_{stor} = A_g$; $h_{stor,air}$ refers to the convection heat transfer coefficient between the heat storage layer surface and the air inside the collector; $Q_{stor,g}$ refers to the radiation heat transfer flux between the heat storage layer surface and the collector surface, according to the analysis above, we have the equation as follows: $Q_{stor,g} = -Q_{g,stor} = A_g\sigma\left(T_{stor}^4 - T_g^4\right)$. $Q_{stor,down}$ refers to the heat transfer flux from the heat storage layer surface to the heat storage layer: $Q_{stor,down} = -A_{stor}\lambda_{stor}\frac{dT}{dx}$, in which λ_{stor} refers to the heat conductivity of the heat storage layer. τ refers to the collector material's solar radiation transmissivity; η refers to the heat storage layer surface's solar radiation absorptivity.

3. Balance conditions for chimney surface

As taking the heat dissipation from chimney to the ambience into account will make the simulation more practical, the third boundary condition is set as follows:

$$-\lambda\left(\frac{\partial t}{\partial y}\right)_w = h(T_w - T_{enviro}) \tag{4.24}$$

In the equation above, the convection heat transfer coefficient is as follows: $h = 5.7 + 3.8V_{enviro}$, where V_{enviro} refers to the ambient wind velocity, and T_{enviro} refer to the ambient temperature. However, considering the facts that the heat convection area is far smaller than that of the collector canopy and the chimney wall is rather thick, it can be considered as adiabatic.

4. Conditions for collector inlet

In accordance with Refs. [13,14], it is assumed that the pressure at the collector inlet is the same as the ambience, hence the collector inlet pressure condition can be set as follows:

$$p_{r,inlet} = 0 \tag{4.25}$$

The $k - \varepsilon$ model is applied for the fluid flow within the system, the standard wall function is applied for wall processing, the SIMPLE algorithm is applied for pressure-velocity coupling, and the QUICK format is applied for equations including momentum equation, energy equation, and other equations. When the mesh number exceeds 120 million, the computation results is grid-independent.

4.6 VALIDITION

Comparison between the numerical simulation results and the experimental results collected on Sep. 2, 1982 of the Spanish SC system is now carried out in order to verify the validity of the numerical simulation results in this chapter. The computation parameters are set according to literature [2], and solar radiation and ambient parameters are set as key conditions for the numerical simulation.

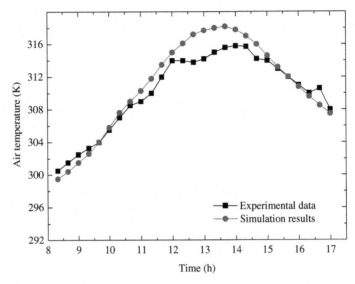

FIGURE 4.16

Comparison between numerical simulation results and experimental results.

As indicated in Fig. 4.16, the simulation results are quite consistent with the experimental results. In detail, the simulation results fluctuate slightly but its curve is relatively quite smooth. Moreover, its difference from the experimental results is less than 5%. The simulation results are slightly higher than experimental results at noon, which is because the ambient wind velocity was not measured, whereas when we consider the collector canopy's heat dissipation we assume that the ambient wind velocity is just 2 m/s, which may have underestimated the heat dissipation from the system to the ambience. Hence it can be concluded that the numerical simulation method applied in this paper is effective and feasible.

4.7 COMPUTATION RESULTS AND ANALYSIS

4.7.1 COMPARISON ON FLOW AND HEAT TRANSFER CHARACTERISTICS

Fig. 4.17 shows the comparison of relative static pressure distributions between the Spanish SC prototype plant and the helical heat-collecting SC system. As shown in this figure, the relative static pressure distribution within the chimney of the new type SC system experiences remarkable variations after applying the helical heat-collecting structure for the collector. The biggest relative static pressure difference at the chimney bottom between the two models is about 15 Pa, which illustrates that the Spanish SC prototype plant is a bit higher than helical

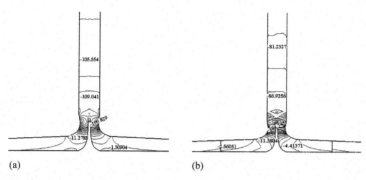

FIGURE 4.17

Comparison of relative static pressure distributions for the two type models (Pa). (a) Model of Spanish SC System. (b) Model of helical heat-collecting SC System.

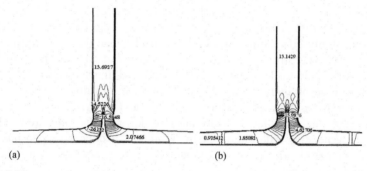

FIGURE 4.18

Comparison of velocity distributions for the two models (m/s). (a) Model of Spanish SC System. (b) Model of helical heat-collecting SC System.

heat-collecting SC system. But on the other hand, at a height of 20 m of the chimney, the relative static pressures of Spanish SC prototype plant and helical heat-collecting SC system are −109 and −86 Pa, respectively, which illustrates that the pressure variation within the helical heat-collecting SC system is more remarkable. As the turbine of the SC system is pressure based, the bigger the pressure variation within the SC system, the bigger the pressure head provided for the turbine. Therefore, as shown in Fig. 4.17b, the helical heat-collecting SC system is able to provide the wind turbine with a bigger pressure head, which is positive for the improvement of the generation power output and the energy conversion efficiency of the whole SC system.

Fig. 4.18 shows the comparison of velocity distributions between the Spanish SC prototype plant and the helical heat-collecting SC system. It can be seen from this figure that the velocity distributions inside the chimney of the two types of models are rather close to each other in despite of slight differences. At the same

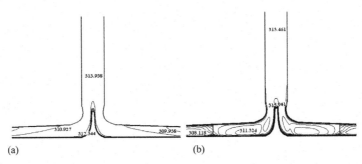

(a)   (b)

FIGURE 4.19

Comparison of temperature distributions for the two models (K). (a) Model of Spanish SC System. (b) Model of helical heat-collecting SC System.

height of the chimney, the velocity of the Spanish SC prototype plant is a little higher than that of the helical heat-collecting SC system, for which the two main reasons are as follows: firstly, the collector radius of the helical heat-collecting SC system is only 90 m, which obviously is smaller than that of the Spanish SC prototype plant (122 m); secondly, although applying the helical heat-collecting type drives the fluid to flow in a helical route toward the chimney and the actual length of flow accordingly reaches 145 m, the helical flow direction keeps changing all the way resulting in an increase of the fluid flow resistance, which not only decreases the fluid flow velocity but also decreases the pressure head used for turbine generating within the system. Therefore, it is necessary to take flow resistance into account and avoid its excessive increase when designing the helical heat-collecting walls.

Fig. 4.19 shows the comparison of temperature distributions between the Spanish SC prototype plant and helical heat-collecting SC system. As shown, the temperature distributions of the two types of models at the same height of the chimney are also very close to each other, which illustrates that the heat absorbed by the fluid during the flow and the according temperature rise of the helical heat-collecting SC system are both close to the Spanish SC prototype plant. Nevertheless, it is found through further observance that the temperature distributions within the collector of the two systems are remarkably different from each other, for which the reason is that the fluid flow direction within the helical heat-collecting SC system is not radial but helical, hence the temperature step change shown within the collector is mainly a result of the difference of the channel of fluid flow.

4.7.2 COMPARISON OF OUTPUT POWER FOR THE TWO TYPE OF MODELS

For further comparison on output power and generating characteristics between the two types of models, the output power under different solar radiations and

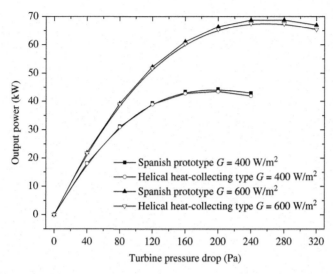

FIGURE 4.20

Comparison of output power for the two models.

turbine pressure drops based upon the above mentioned two models are examined. The turbine of SC system is pressure-based, the turbine efficiency is set as 72%, and the output power is the product of the system volume flow, turbine pressure drop, and the turbine efficiency. Fig. 4.20 shows the output power computation results comparison between the Spanish SC prototype plant and the helical heat-collecting SC system, herein G refers to the solar radiation. As shown in the figure, the output powers of the two models are very close to each other under the same solar radiation and turbine pressure drop, and the biggest deviation is less than 3%. Thus, it can be concluded that the helical heat-collecting SC system is of better economical efficiency and bigger commercial advantage when compared with the traditional SC system.

Attention must be paid to the fact that it is not necessary to compare the numerical simulation results with the experimental results of the Spanish SC prototype plant, as the turbine efficiency of the Spanish SC prototype plant was not well designed or optimized. The efficiencies of new turbines specially designed for SC systems have all surpassed 72% [10,11,23], therefore, it is reasonable to set 72% as the turbine efficiency while comparing the numerical results of the two types of SC plants.

4.7.3 COMPARISON OF DIFFERENT HELICAL-WALL SC SYSTEMS

Figs. 4.21 and 4.22 shows different models of helical-wall SC systems and the corresponding output powers when the solar radiation is 600 W/m^2. From Fig. 4.22 we

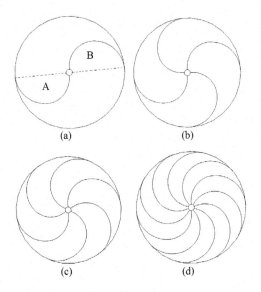

FIGURE 4.21

Different models of helical-wall SC systems: (a) 2-helical-wall SC system; (b) 4-helical-wall SC system; (c) 6-helical-wall SC system; (d) 10-helical-wall SC system.

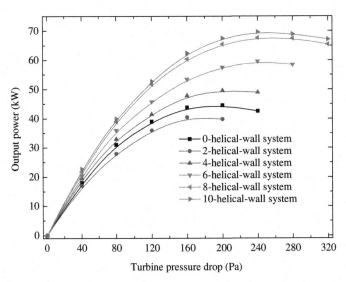

FIGURE 4.22

Comparison of output power with different helical-wall SC systems when $G = 600$ W/m^2.

can see that the number of helical walls have a significant influence on the output power of the SC system. With the same turbine pressure drop, the output powers of system with 4 helical walls are a little higher than that of systems without helical wall, whereas the output powers of systems with 2 helical walls are a little lower than that of systems without helical walls. This is because the helical fluid flow inside the collector does not form if the system has less than 4 helical walls. Furthermore, in the A and B regions shown in Fig. 4.21a, it is due to the existence of the 2 helical walls of the system that the fluid flow through the chimney is blocked and the output power decreases. As is shown in Fig. 4.22, when the SC system has 6 helical walls or more, the output power increases greatly, the helical fluid flow forms, and the system will operate in a good condition. Apparently, setting up 8 helical walls to this small SC system is the optimal selection.

4.7.4 CONTRAST ON COLLECTOR'S INITIAL INVESTMENT

Table 4.2 shows the comparison of the collector's initial investment and output power with different types of SC system. The collector area of the Spanish SC prototype plant with a radius of 122 m, is 46,681 m^2, excluding the area of the chimney. In contrast, the collector area of the helical heat-collecting SC system with a radius of 90 m is 25,368 m^2, with the collector area reduced by 45.7% (reduced area is 21,313 m^2). The height of the helical transparent wall gradually increases from 2 m at the inlet to 6 m at the center, thus the average height of the helical wall can be assumed as 4 m, and then the total area of the 2 newly added helical transparent walls is 1160 m^2. The helical transparent wall mainly serves as a rotational flow creator within the collector and has no extra requirements on architecture, hence it is feasible to assume that its construction cost is the same as that of the collector canopy. Therefore, the collector area of helical heat-collecting SC system is the sum of the collector canopy area and the helical walls area. In terms of the initial investment on the collector, the five helical heat-collecting SC systems are lower than the Spanish SC prototype plant by about 33.2–43.2%.

Table 4.2 Comparison of Investment and Output Power With Different Types of SC System

SC Systems	R_{coll} (m^2)	A_{coll} (m^2)	A_{wall} (m^2)	A_{tot} (m^2)	A_{reduc} (m^2)	I_{reduc} (%)
Spanish SC	122	46,681	0	46,681	–	–
0-helical-wall	90	25,368	0	25,368	21,313	45.7
2-helical-wall	90	25,368	1160	26,528	20,153	43.2
4-helical-wall	90	25,368	2320	27,688	18,993	40.7
6-helical-wall	90	25,368	3480	28,848	17,833	38.2
8-helical-wall	90	25,368	4640	30,008	16,673	35.7
10-helical-wall	90	25,368	5800	31,168	15,513	33.2

As indicated by the computation results, when the collector radius is kept constant at 90 m, the computation results of the two models are very close to each other without notably changing the system's generation characteristics or fluid flow and heat transfer characteristics when 8 helical heat-collecting walls are applied; in contrast, applying just 2 helical heat-collecting walls is hardly able to drive the rotational flow within the collector to flow into the chimney. In addition, although the output power of the helical heat-collecting SC system is nearly the same as the Spanish SC prototype plant under the same solar radiation during the day, it is possible for the helical heat-collecting SC system's generation ability to decrease remarkably if we consider the system's day-night generation characteristics where the new system actually decreases the heat storage when it decreases the collector area, probably resulting in a remarkable decrease of generation ability at night, which especially should be paid attention to during the designing and application of the SC system.

China has started research on SC technology and the feasibility reasoning of large-scale commercial SC power plant. However, no SC system is established up to now, for which the main reason is the excessive initial investment. If the helical heat-collecting SC system is applied, the initial investment of the system can be greatly reduced, making it possible to establish a demonstration commercial SC power plant with the output power of $5 \sim 10$ MW in China.

4.8 CONCLUSION

Two-dimensional steady-state numerical simulations for the solar chimney power plant system, which includes collector, chimney, turbine, and energy storage layer, are carried out in this paper. The turbine is regarded as a pressure-based one which is different from that in Ref. [13] and the energy storage layer is treated as a porous media, as mentioned in Ref. [15], aiming to analyze the pressure, velocity, and temperature distributions of the system. Meanwhile, the influence of the turbine pressure drop on the flow and heat transfer characteristics and the output power of the SC system are considered. The influences of solar radiation and turbine efficiency on the output power and energy loss of the system are also analyzed. Through analysis, it is found that the influences of solar radiation and pressure drop across the turbine are rather considerable; large outflow of heated fluid from the chimney outlet becomes the main cause for the energy loss of the system, and the canopy also causes considerable energy loss.

A new type of helical heat-collecting SC system has been designed. Through numerical simulation on its flow and heat transfer characteristics, it is found that, considering the fact that the collector radius of the helical heat-collecting SC system is just 90 m, the new model's radius is decreased by 25%, while its land area and materials are decreased by about $33 \sim 43\%$ when compared with the Spanish

SC prototype plant. The differences of fluid flow and heat transfer characteristic parameters between the Spanish SC prototype plant and the 8-helical-wall heat-collecting SC system are slight. Therefore, it can be concluded that helical heat-collecting SC system is of better economical efficiency and bigger commercial advantage than the traditional SC system, especially when we take the initial investment into account.

NOMENCLATURE

u	Average velocity magnitude in the axial direction (m/s)
r	Coordinate in radial direction (m)
T	Temperature (K)
T_0	Temperature of the environment (K)
g	Gravity, $g = 9.807$ (m/s^2)
Pr	Prandtl number (dimensionless), $\text{Pr} = \frac{c_p \mu}{k}$
x	Coordinate in axial direction (m)
S_ϕ	Source item

GREEK SYMBOLS

v	Kinematic viscosity (m^2/s)
μ_t	Turbulent dynamic viscosity coefficient
μ	Dynamic viscosity (kg/(m·s))
β	Coefficient of cubic expansion
ρ	Density (kg/m^3)
κ	Turbulent kinetic energy (J/kg)
ε	Dissipation (W/kg)
σ_ε	Prandtl number of dissipation
σ_T	Prandtl number of turbulence
σ_κ	Prandtl number of pulse kinetic energy 1.0
Γ_ϕ	Generalized diffusion coefficient

SUBSCRIPT

e	Environment
t	Viscous dissipation effect caused by turbulent characteristics
ε	Dissipation
κ	Kinetic energy
ϕ	Porosity
T	Turbulent flow

REFERENCES

[1] Haaf W, Friedrich K, Mayer G, Schlaich J. Solar chimneys. Int J Solar Energy 1983;2:3−20.

[2] Haaf W, Friedrich K, Mayer G, Schlaich J. Solar chimneys. Int J Solar Energy 1984;2:141−61.

[3] Krisst R. Energy transfer system. Altern Sources Energy 1983;63:8−11.

[4] Kulunk H, A prototype solar convection chimney operated under izmit condition. In: Veziroglu TN, editor. Prod. 7th MICAS, Hemisphere Publishing Corporation, Miami Beach, FL, USA 1985, p. 162.

[5] Pasumarthi N, Sherif SA. Experimental and theoretical performance of a demonstration solar chimney model—part I: mathematical model development. Int J Energ Res 1998;22:277−88.

[6] Bernardes MAD, Valle RM, Cortez MFB. Numerical analysis of natural laminar convection in a radial solar heater. Int J Therm Sci 1999;38:42−50.

[7] Bilgen E, Rheault J. Solar chimney power plants for high latitudes. Sol Energy 2005;79:449−58.

[8] Pretorius JP, Kroger DG. Solar chimney power plant performance. J Sol Energ-T Asme 2006;128:302−11.

[9] Gannon AJ, von Backstrom TW. Solar chimney cycle analysis with system loss and solar collector performance. J Sol Energ-T Asme 2000;122:133−7.

[10] Gannon AJ, von Backstrom TW. Solar chimney turbine performance. J Sol Energ-T Asme 2003;125:101−6.

[11] von Backstrom TW, Gannon AJ. Solar chimney turbine characteristics. Sol Energy 2004;76:235−41.

[12] Bernardes MAD, Voss A, Weinrebe G. Thermal and technical analyses of solar chimneys. Sol Energy 2003;75:511−24.

[13] Pastohr H, Kornadt O, Gurlebeck K. Numerical and analytical calculations of the temperature and flow field in the upwind power plant. Int J Energ Res 2004;28:495−510.

[14] Ming TZ, Wei L, Xu GL. Analytical and numerical investigation of the solar chimney power plant systems. Int J Energ Res 2006;30:861−73.

[15] Ming TZ, Liu W, Pan Y, Xu GL. Numerical analysis of flow and heat transfer characteristics in solar chimney power plants with energy storage layer. Energ Convers Manage 2008;49:2872−9.

[16] Liu W, Ming TZ, Yang K, Pan Y. Simulation of characteristic of heat transfer and flow for MW-graded solar chimney power plant system. J Huazhong Univ Sci Technolog 2005;33:5−7.

[17] Ming TZ, Liu W, Xu GL, Xiong YB, Guan XH, Pan Y. Numerical simulation of the solar chimney power plant systems coupled with turbine. Renew Energ 2008;33:897−905.

[18] Kuznetsov AV. Analytical investigation of heat transfer in Couette flow through a porous medium utilizing the Brinkman-Forchheimer-extended Darcy model. Acta Mech 1998;129:13−24.

[19] Papageorgiou CD. External wind effects on floating solar chimney. In: Proceedings of the fourth IASTED international conference on power and energy systems 2004. p. 159−63.

[20] Papageorgiou CD. Floating solar chimney power stations with thermal storage. In: Proceedings of the sixth IASTED international conference on European power and energy systems 2006. p. 325–31.

[21] Papageorgiou C. Floating solar chimney technology: a solar proposal for China. Proc Ises Solar World Congr 2007: Solar Energy Human Settlement 2007;Vols I-V:172–6.

[22] Papageorgiou CD, Katopodis P. A modular solar collector for desert floating solar chimney technology. Energy Environ Eng S 2009:126–32.

[23] Denantes F, Bilgen E. Counter-rotating turbines for solar chimney power plants. Renew Energ 2006;31:1873–91.

Design and simulation method for SUPPS turbines*

Tingzhen Ming[1,2], Wei Liu[2], Guoliang Xu[2], Yanbin Xiong[2], Xuhu Guan[2] and Yuan Pan[3]

[1]*School of Civil Engineering and Architecture, Wuhan University of Technology, Wuhan, P.R. China* [2]*School of Energy and Power Engineering, Huazhong University of Science and Technology, Wuhan, P.R. China* [3]*School of Electrical and Electric Engineering, Huazhong University of Science and Technology, Wuhan, P.R. China*

CHAPTER OUTLINE

5.1 INTRODUCTION

The solar chimney power plant system (SCPPS), which consists of four major components: the collector, the chimney, the turbine, and the energy storage layer, was first proposed in the late 1970s by Professor Jörg Schlaich and tested with a prototype model in Manzanares, Spain in the early 1980s [1]. Air underneath the low circular transparent glass or film canopy open at the circumference is heated by radiation from the sun. The canopy and the surface of the energy storage layer

---

*The content of this chapter was published in Renewable Energy

Solar Chimney Power Plant Generating Technology. DOI: http://dx.doi.org/10.1016/B978-0-12-805370-6.00005-3

FIGURE 5.1

Schematic drawing of a solar chimney.

below form an energy collection system, called a collector. The chimney, a vertical tower tube with large air inlets at its base, stands in the center of the collector (Fig. 5.1). The joint between the collector and the chimney is airtight. The wind turbine is installed at the bottom of the chimney where there is a large pressure difference from the outside. For the large-scale solar chimney systems, there may be several wind turbines inside as it is difficult, at the current time, to produce wind turbines with a rated load over 10 MW. As the density of hot air inside the system is less than that of the cold air in the environment at the same height, natural convection affected by buoyancy which acts as a driving force comes into existence. The energy of the air flow is converted into mechanical energy by pressure-staged wind turbines at the base of the tower, and ultimately into electrical energy by electric generators coupled to the turbines.

As the SCPPSs could make significant contributions to the energy supplies of those countries where there is plenty of desert land, which is not being utilized, and sunlight in Africa, Asia, and Oceania, researchers have made many reports on this technology in the recent few decades. Haaf et al. provided fundamental investigations for the Spanish prototype system in which the energy balance, design criteria, and cost analysis were discussed [1]. The next year, the same authors reported preliminary test results of the solar chimney power plant [2]. Krisst demonstrated a "back yard type" device with a power output of 10 W in West Hartford, Connecticut, USA [3]. Kulunk produced a microscale electric power plant of 0.14 W in Izmit, Turkey [4]. Pasumarthi and Sherif developed a mathematical model to study the effect of various environment conditions and geometry on the air temperature, air velocity, and power output of the solar chimney [5]. Pasumarthi and Sherif also developed three model solar chimneys in Florida and reported the experimental data to assess the viability of the solar chimney concept

[6]. Padki and Sherif developed a simple model to analyze the performance of the solar chimney [7]. Lodhi presented a comprehensive analysis of the chimney effect, power production, efficiency, and estimated the cost of the solar chimney power plant set up in developing nations [8]. Bernardes et al. presented a theoretical analysis of a solar chimney, operating on natural laminar convection in steady state [9]. Gannon and Backström presented an air standard cycle analysis of the solar chimney power plant for the calculation of limiting performance, efficiency, and the relationship between the main variables, including chimney friction, system, turbine, and exit kinetic energy losses [10]. Gannon and Backström presented an experimental investigation of the performance of a turbine for the solar chimney systems, the measured results showed that the solar chimney turbine presented has a total-to-total efficiency of 85−90% and a total-to-static efficiency of 77−80% over the design range [11]. Later, Backström and Gannon presented analytical equations in terms of turbine flow and load coefficient and degree of reaction to express the influence of each coefficient on turbine efficiency [12]. Bernardes et al. developed a comprehensive thermal and technical analysis to estimate the power output and examine the effect of various ambient conditions and structural dimensions on the power output [13]. Pastohr et al. carried out a numerical simulation to improve the description of the operation mode and efficiency by coupling all parts of the solar chimney power plant including the ground, collector, chimney, and turbine [14]. Schlaich et al. presented theory, practical experience, and economy of solar chimney power plant to give a guide for the design of 200 MW commercial SCPPSs [15]. Ming et al. presented a thermodynamic analysis on the solar chimney power plant and advanced energy utilization degree to analyze the performance of the system which can produce electricity day and night [16]. Liu et al. carried out a numerical simulation for the MW-graded solar chimney power plant, presenting the influences of pressure drop across the turbine on the draft and the power output of the system [17]. Bilgen and Rheault designed a solar chimney system for power production at high latitudes and evaluated its performance [18]. Pretorius and Kröger evaluated the influence of a developed convective heat transfer equation, more accurate turbine inlet loss coefficient, quality of collector roof glass, and various types of soil on the performance of a large-scale solar chimney power plant [19]. Ming et al. developed a comprehensive model to evaluate the performance of a SCPPS, in which the effects of various parameters on the relative static pressure, driving force, power output, and efficiency have been further investigated [20]. Zhou et al. presented experimental and simulation results of a solar chimney thermal power generating equipment in China, and based on the simulation and the specific construction costs at a specific site, the optimum combination of chimney and collector dimensions was selected for a required electric power output [21].

For a SCPPS with certain geometrical dimensions, numerical simulations with no load condition were presented in previous reports [17]. Numerical simulation and analysis of a 2-D SCPPS including the turbine and the energy storage layer were explored based on the numerical CFD program FLUENT [14], but the turbine was regarded as a reverse fan and the pressure drop across the turbine was given by the Beetz power limit: $\Delta p_t = -\frac{8}{27}\rho u^2$, so the simulations did not give results of the power output, turbine efficiency, and chimney outlet parameters on the turbine rotational speed. In the present work, the authors will show a preliminary investigation on the 3-D SCPPSs coupled with the turbine in order to explore the problems as described, and will also present a mathematical model for the turbine region and simulation results for the MW-graded SCPPS.

5.2 NUMERICAL MODELS

It might be a little difficult to carry out the numerical simulations on the SCPPSs coupled with the collector, chimney, and turbine. In this work, the main dimensions of the physical model shown in Table 5.1 were selected according to the Spanish prototype [1], with a 3-blade or 4-blade pressure-staged turbine with CLARK aerofoil installed at the bottom of the chimney. Tables 5.2 and 5.3 show in detail the parameters of the aerofoil type and blade section. Moreover, numerical simulation on the MW-graded SCPPS coupled with a 5-blade pressure-staged turbine with CLARK aerofoil has also been carried out, in which the physical model of the turbine is shown in Fig. 5.2. The energy storage layer is not included in the SCPPS because it is not necessary to take the energy storage layer into consideration when carrying out the steady state numerical simulation.

Table 5.1 Main Dimensions of the Solar Chimney Models

		Spanish Prototype	MW-Graded Model
Collector	Radius (m)	122	1500
	Height (m)	2–6 (from inlet to center)	4–8 (from inlet to center)
Chimney	Radius (m)	5	30
	Height (m)	200	400

Table 5.2 Parameters of CLARK Aerofoil Type

Level (%)	0.00	2.65	10.3	22.2	37.1	53.3	69.1	83.0	93.3	99.0	100
Upper (%)	0.00	3.40	6.55	8.78	9.33	8.46	6.40	3.89	1.65	0.25	0.00
Lower (%)	0.00	1.97	2.74	2.72	2.22	1.67	1.12	0.47	0.29	0.053	0.00

Table 5.3 Parameters of Blade Section

r/R	0.2	0.4	0.6	0.8	1.0
Chord (m)	0.9762	0.8572	0.7382	0.6192	0.5000
Stagger angle (°)	9.524	7.144	4.764	2.384	0.000

FIGURE 5.2

Turbine model of the MW-graded solar chimney power plant.

5.3 MATHEMATICAL MODELS

5.3.1 IN THE COLLECTOR AND CHIMNEY REGIONS

Some assumptions should be taken into account to simplify the problem to get the performance of the SCPPS under load condition. The whole system should be divided into three regions: the collector, the turbine, and the chimney. For the collector and chimney regions, conventional methods of numerical simulation could be used to give reasonable results, while air flow in the turbine region should be characterized by the control equations in a rotating reference frame.

For the natural convection system, it might be necessary to take into account the value of Ra Number, which can be used as a criteria for laminar and turbulent flow in the system.

$$Ra = \frac{g\beta(T_h - T_c)L^3}{\alpha\nu} \tag{5.1}$$

where T_h, T_c are the highest and the lowest temperature of the system, respectively, and L is the height of the collector. It shows, by simple analysis, that the Ra Number of the SCPPS is higher than the critical Ra number, 10^9, which means that turbulent flow happens in almost the whole system except for the entrance of the collector inlet. So the control equations including the continuity equation, momentum equation, energy equation, and the standard $k - \varepsilon$ equation in the collector and chimney regions can be written as follows:

Continuity equation:

$$\frac{\partial\rho}{\partial t} + \frac{\partial(\rho u)}{\partial x} + \frac{\partial(\rho v)}{\partial y} + \frac{\partial(\rho w)}{\partial z} = 0 \tag{5.2}$$

Momentum equations:

$$\frac{\partial(\rho u)}{\partial t} + \frac{\partial(\rho uu)}{\partial x} + \frac{\partial(\rho vu)}{\partial y} + \frac{\partial(\rho wu)}{\partial z} = -\frac{\partial p}{\partial x} + \mu\left(\frac{\partial^2 u}{\partial x^2} + \frac{\partial^2 u}{\partial y^2} + \frac{\partial^2 u}{\partial z^2}\right) \tag{5.3}$$

$$\frac{\partial(\rho v)}{\partial t} + \frac{\partial(\rho uv)}{\partial x} + \frac{\partial(\rho vv)}{\partial y} + \frac{\partial(\rho wv)}{\partial z} = -\frac{\partial p}{\partial y} + \mu\left(\frac{\partial^2 v}{\partial x^2} + \frac{\partial^2 v}{\partial y^2} + \frac{\partial^2 v}{\partial z^2}\right) \tag{5.4}$$

$$\frac{\partial(\rho w)}{\partial t} + \frac{\partial(\rho uw)}{\partial x} + \frac{\partial(\rho vw)}{\partial y} + \frac{\partial(\rho ww)}{\partial z} = \rho g\beta(T - T_\infty) + \mu\left(\frac{\partial^2 w}{\partial x^2} + \frac{\partial^2 w}{\partial y^2} + \frac{\partial^2 w}{\partial z^2}\right) \tag{5.5}$$

Energy equation:

$$\frac{\partial(\rho cT)}{\partial t} + \frac{\partial(\rho cuT)}{\partial x} + \frac{\partial(\rho cvT)}{\partial y} + \frac{\partial(\rho cwT)}{\partial z} = \lambda\left(\frac{\partial^2 T}{\partial x^2} + \frac{\partial^2 T}{\partial y^2} + \frac{\partial^2 T}{\partial z^2}\right) \tag{5.6}$$

k and ε equations:

$$\frac{\partial}{\partial t}(\rho k) + \frac{\partial}{\partial x_i}(\rho k u_i) = \frac{\partial}{\partial x_j}\left(\left(\mu + \frac{\mu_t}{\sigma_k}\right)\frac{\partial k}{\partial x_j}\right) + G_k + G_b - \rho\varepsilon + S_k \tag{5.7}$$

$$\frac{\partial}{\partial t}(\rho\varepsilon) + \frac{\partial}{\partial x_i}(\rho\varepsilon u_i) = \frac{\partial}{\partial x_j}\left(\left(\mu + \frac{\mu_t}{\sigma_\varepsilon}\right)\frac{\partial\varepsilon}{\partial x_j}\right) + C_{1\varepsilon}(G_k + C_{3\varepsilon}G_b) - C_{2\varepsilon}\rho\frac{\varepsilon^2}{k} + S_\varepsilon \tag{5.8}$$

In the equations above, G_k represents the generation of turbulence kinetic energy due to the mean velocity gradients, and G_b is the generation of turbulence kinetic energy due to buoyancy. For the standard $k - \varepsilon$ models, the constants have the following values [22]:

$$C_{1\varepsilon} = 1.44, \quad C_{2\varepsilon} = 1.92, \quad C_\mu = 0.09, \quad \sigma_k = 1.0, \quad \sigma_\varepsilon = 1.3.$$

5.3.2 IN THE TURBINE REGION

When successfully creating models of the collector and chimney regions using Gambit, we will typically analyze the fluid flow in an inertial reference frame, that is, in a nonaccelerating coordinate system using FLUENT. However, the air flow in the turbine region is different from that in the collector and the chimney regions, and it also rotates with the blades at a certain velocity in the radial direction besides passing through the whole region, which makes it very difficult for us to carry out the numerical simulation. Fortunately, however, FLUENT has the ability to model flows in a rotating reference frame, in which the acceleration of the coordinate system is included in the equations of motion describing the fluid flow. Such flow as in the turbine region of the SCPPSs can also be modeled in a coordinate system that is moving with the rotating equipment and thus experiences a constant acceleration in the radial direction. When the flow in the turbine region is defined in a rotating reference frame, the rotating boundaries become stationary relative to the rotating frame, because they are moving at the same speed as the reference frame. The turbine-chimney interaction could be treated by applying the multiple reference frame (MRF) model.

Thereby, when the equations of motion are solved in a rotating reference frame, the acceleration of the fluid is augmented by additional terms that appear in the momentum equations. The absolute velocity and the relative velocity are related by the following equation:

$$\vec{v}_r = \vec{v} - \left(\vec{\Omega} \times \vec{r} \right) \tag{5.9}$$

where, $\vec{\Omega}$ and $\vec{r}$ are the angular velocity vector and position vector in the rotating frame, respectively. So the momentum equation for the rotation coordinate system can be obtained as follows:

$$\frac{\partial}{\partial t} \left(\rho \vec{v} \right) + \nabla \cdot \left(\rho \vec{v}_r \vec{v} \right) + \rho \left(\vec{\Omega} \times \vec{r} \right) = \nabla \cdot \left(\mu \nabla \vec{v}_r \right) + S_{\vec{v}_r} \tag{5.10}$$

Substituting Eq. (5.9) into Eq. (5.10) yields:

$$\frac{\partial}{\partial t} \left(\rho \vec{v}_r \right) + \nabla \cdot \left(\rho \vec{v}_r \rightarrow v_r \right) + \rho \left(2\vec{\Omega} \times \vec{r} + \vec{\Omega} \times \vec{\Omega} \times \vec{r} \right) + \rho \frac{\partial \vec{\Omega}}{\partial t} \times \vec{r} = \nabla \cdot \left(\mu \nabla \vec{v}_r \right) + S_{\vec{v}_r} \tag{5.11}$$

where $\rho \left(2\vec{\Omega} \times \vec{r} + \vec{\Omega} \times \vec{\Omega} \times \vec{r} \right)$ is the Coriolis force.

For flows in rotating zones, the continuity equation can be written as follows:

$$\frac{\partial \rho}{\partial t} + \nabla \cdot \left(\rho \cdot \vec{v}_r \right) = 0 \tag{5.12}$$

The maximal available energy of the updraft air in the system could be derived as follows:

$$N = V \Delta p \tag{5.13}$$

where, V and Δp are the mass flow rate and the pressure drop across the turbine, respectively. The power output, or the technical work, from the turbine can be obtained:

$$P_t = \frac{2\pi n I}{60} \tag{5.14}$$

where, n and I are the rotation speed and the total blade moments of the turbine, respectively. From Eqs. (5.13) and (5.14), the turbine efficiency can be obtained:

$$\eta = \frac{P_t}{N} = \frac{2\pi n I}{60 V \Delta p} \tag{5.15}$$

5.4 NEAR-WALL TREATMENTS FOR TURBULENT FLOWS

Turbulent flows are significantly affected by the presence of walls. The near-wall modeling significantly impacts the fidelity of numerical solutions, inasmuch as walls are the main source of mean vorticity and turbulence. In the near-wall zone the solution variables have large gradients, and the momentum and other scalar transports occur most vigorously. Therefore, accurate representation of the flow in the near-wall zone determines successful predictions of wall-bounded turbulent flows.

The wall function approach, which uses semiempirical formulas to bridge the viscosity-affected zone between the wall and the fully-turbulent zone, will be adopted in this model because it can substantially save computational resources, obviate the need to modify the turbulence models to account for the presence of the wall, and because the viscosity-affected near-wall zone, in which the solution variables change most rapidly, does not need to be resolved. In addition, this approach is economical, robust, and reasonably accurate. It is a practical option for the near-wall treatments for industrial flow simulations.

Based on the proposal of Launder and Spalding [23], the law-of-the-wall for mean velocity yields:

$$U^* = \frac{1}{\kappa} \ln \left(EY^* \right) \tag{5.16}$$

where

$$U^* \equiv \frac{U_P C_\mu^{1/4} k_P^{1/2}}{\tau_w / \rho} \tag{5.17}$$

$$Y^* \equiv \frac{\rho C_\mu^{1/4} k_P^{1/2} Y_P}{\mu} \tag{5.18}$$

and U_P, K_P, Y_P are the mean velocity, turbulence kinetic energy of the fluid at point P and distance from point P to the wall, respectively. U^* is the nondimensional velocity and Y^* is the nondimensional distance from the computational point to the wall.

The logarithmic law for mean velocity is employed when $Y^* > 11.225$. When the mesh is such that $Y^* < 11.225$ at the wall-adjacent cells, the laminar stress-strain relationship can be written as $U^* = Y^*$. Reynolds' analogy between momentum and energy transport gives a similar logarithmic law for mean temperature. As in the law-of-the-wall for mean velocity, the law-of-the-wall for temperature employed comprises two different laws: linear law for the thermal conduction sublayer where conduction is important and logarithmic law for the turbulent zone where effects of turbulence dominate conduction shown as follows:

$$T^* = \Pr Y^* + \frac{1}{2} \rho \Pr \frac{C_\mu^{1/4} k_P^{1/2}}{\dot{q}} U_P^2 \quad \left(Y^* < Y_T^* \right) \tag{5.19}$$

$$T^* = \mathrm{Pr}_t(U^* + B) + \frac{1}{2}\rho\frac{C_\mu^{1/4}k_P^{1/2}}{\dot{q}}\left(\mathrm{Pr}_t U_p^2 + \left(\mathrm{Pr} - \mathrm{Pr}_t U_c^2\right)\right) \quad \left(Y^* < Y_T^*\right) \qquad (5.20)$$

Where B is computed by using the formula given by Jayatilleke [24]:

$$B = 9.24\left(\left(\frac{\sigma}{\sigma_t}\right)^{3/4} - 1\right)\left(1 + 0.28e^{-0.007\sigma/\sigma_t}\right) \qquad (5.21)$$

and T_P, T_W are temperature at the cell adjacent to wall and temperature at the wall, respectively, U_C is mean velocity magnitude at $Y^* = Y^*_T$.

The nondimensional thermal sublayer thickness, Y_T^*, in Eqs. (5.19) and (5.20) is computed as the Y^* value at which the linear law and the logarithmic law intersect, given the Pr value of the water being modeled.

5.5 NUMERICAL SIMULATION METHOD

The collector inlet condition is variable under different solar radiation as the free convection happens in the whole system. Thereby, some important parameters such as the collector inlet velocity, mass flow rate of the system, pressure drop across the turbine, and the correlation between the solar radiation and the rotational speed of the turbine are unknown for the numerical simulation of the system including the collector, turbine, and chimney regions. Fortunately, however, different rotational speed of the turbine could be given in advance without taking into account the variation of the solar radiation intensity, which implies that not much more attention need be paid to the correlation between the solar radiation and the rotational speed of the turbine.

It is necessary to point out that the rotational direction of the turbine in the solar chimney might have significant effect on the validity of the results of numerical simulation. If the fluid impels the turbine to rotate, which transforms the energy of the air with relative higher temperature and velocity into technical work, the mass flow rate of the system and chimney outlet velocity will decrease compared with those on no load condition. While the turbine blades will impel the air flows out of the chimney, if the rotational speed of the turbine is set in an opposite direction, therefore, the mass flow rate of the system and the chimney outlet velocity will increase inversely.

Numerical simulations of the SCPPSs should be based on some assumptions shown as follows: (1) Constant environment conditions, including uniform solar radiation (800 W/m²), ambient temperature, and inlet air temperature, should be taken into account. (2) Axisymmetric air flow in the collector inlet, that is, non-uniform heating of the collector surface in terms of the sun's altitude angle is neglected. (3) Heat loss through the wall of the chimney is neglected. (4) The Boussinesq approximation is assumed to be valid for the variation of air density in the whole system.

Table 5.4 Boundary Conditions and Model Parameters

Place	Type	Value
Surface of the ground	Wall	$T = f_1(r)$ K
Surface of the canopy	Wall	$T = f_2(r)$ K
Surface of the chimney	Wall	$q_{chim} = 0$ W/m^2
Inlet of the collector	Pressure inlet	$p_{r,i} = 0$ Pa, $T_0 = 293$K
Outlet of the chimney	Pressure outlet	$p_{r,o} = 0$ Pa
Turbine rotational speed		$\omega = n$ rpm

The main boundary conditions, derived by energy equilibrium equations, are shown in Table 5.4 for mass, momentum, and energy conservation equations, and the temperature profiles of the ground and the canopy in the collector could be different parabolic functions of the collector radius by taking into account the axisymmetric air flow in the collector shown in the second assumption above, and the functions will vary with different solar radiations.

As mentioned above, both stationary and moving regions are selected for the numerical simulation on the SCPPS coupled with the turbine. Three different models, that is, the MRF model, the mixing plane model, and the sliding mesh model, could be used to solve the problem involved in FLUENT, of which the MRF model is the simplest. It is a steady-state approximation in which individual cell zones move at different rotational speeds. *Interior* faces could be selected for the boundaries (ie, the collector and the turbine regions, the turbine and the chimney regions) between reference frames. In addition, simulation results independent of the meshes should be taken into consideration.

5.6 RESULTS AND DISCUSSIONS

5.6.1 VALIDITY OF THE METHOD FOR THE SPANISH PROTOTYPE

To validate the simulation method used by the author in this chapter, numerical simulation results are compared with the experimental result [2] for the Spanish prototype with a 4-blade turbine which is set up at the bottom of the chimney.

When the solar radiation is 800 W/m^2, 35 kW technical work could be extracted from the Spanish prototype with a 4-blade turbine [2]. Fig. 5.3 shows the simulation results of the SCPPS with a 4-blade designed in this chapter and Fig. 5.4 shows the velocity vectors in detail in the turbine region. It can be easily seen from Fig. 5.3 that the experimental result is in the scope of the simulation results, and the maximum power output from the simulation results is a little higher than the experimental result, and that the rotational speed of the turbine is 80 rpm, 25% less than that of the experimental result [2]. In addition, as velocity changes greatly around the turbine blades, especially at the tip of the blade, air velocity might be a few times larger than the main flow in the chimney.

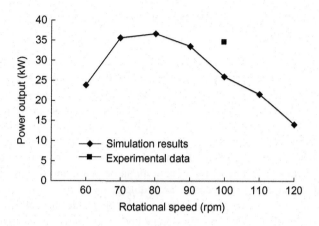

FIGURE 5.3

Comparison between simulation results and experiment data.

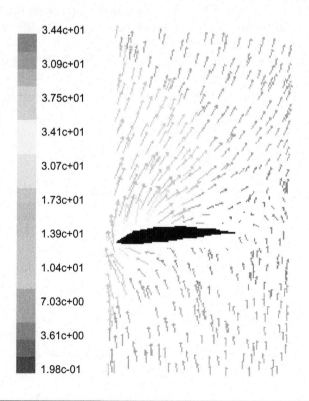

FIGURE 5.4

Velocity vectors in the turbine region.

5.6.2 CHARACTERISTIC OF 3-BLADE TURBINE FOR THE SPANISH PROTOTYPE

Fig. 5.5 shows the effect of different turbine rotational speeds on the average temperature and velocity of the chimney outlet. From the figure, we can see that the average velocity of the chimney outlet decreases significantly and the average temperature inversely with the increase of the turbine rotational speed. The reason is that, when all the other parameters such as the environment parameters and the solar radiation are constant, large resistant force caused by the blades of the turbine occurs with the increase of the turbine rotational speed. And with the increase of the resistant force, the mass flow rate of the system might decrease and then the air velocity of the chimney outlet would decrease accordingly. In addition, the decrease of the air velocity might result in a longer time for the air inside the collector to absorb energy from the surface of the energy storage layer while passing through the collector. Thereby, the average temperature of the chimney outlet increases prominently.

Fig. 5.6 shows the effect of the turbine rotational speed on the pressure drop across the turbine and the mass flow rate of the Spanish prototype. When the turbine rotational speed is lower than 50 rpm, the mass flow rate is higher than 1000 kg/s. The mass flow rate decreases almost linearly with the increase of the turbine rotation speed, which indicates that the resistant force changes significantly with the increase of the turbine rotational speed. It can be predicted that the mass flow rate might decrease more notably when the blades number of the turbine increases.

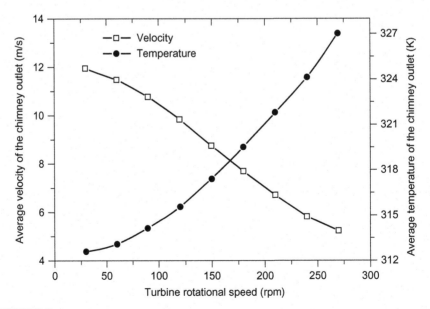

FIGURE 5.5

The effect of rotational speed on the outlet parameters of the chimney.

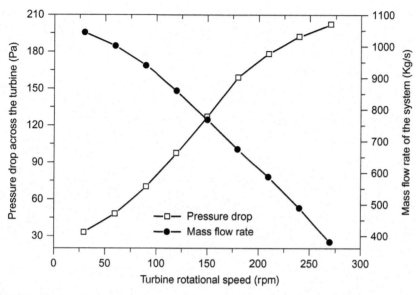

FIGURE 5.6

Pressure drop of the turbine and mass flow rate of the prototype.

Besides, we can see from Fig. 5.6 that the pressure drop across the turbine increases remarkably with the increase of the turbine rotation speed. The increase of pressure drop across the turbine means that an increasing part of the driving force of the SCPPSs is used to drive the turbine. The variation trend of pressure drop with the turbine rotation speed decreases when the turbine rotation speed surpasses 200 rpm, which means that the utilization of pressure drop of the system reaches its limitation and it is not necessary to increase the turbine rotation speed any more.

Fig. 5.7 shows the relationship between the turbine rotational speed and the maximum available energy, power output, and turbine efficiency. For the kW-graded turbine model designed in this work, the maximum available energy is a little above 100 kW when the turbine rotational speed reaches 180 rpm, while the power output and turbine efficiency reach their peak values when the rotational speed reaches 210 rpm. The maximum power output is about 50 kW, which is identical to the designed data of the Spanish prototype, and the turbine efficiency is near 50%, which could reach a much higher value with some optimization for the turbine.

Compared with the Spanish prototype turbine designed with 4 blades, the turbine rotational speed of the 3-blade turbine in this work is much larger, the reasons are as follows: firstly, the number of turbine blades is a significant factor to the rotational speed which will notably increase once the blade number decreases with the same power output, and simulation results show that power output of 50 kW could not be extracted from the solar chimney power plant prototype if a 2-blade turbine is selected to be installed at the bottom of the chimney. Secondly, the turbine blades are a little longer and slimmer to some extent, and

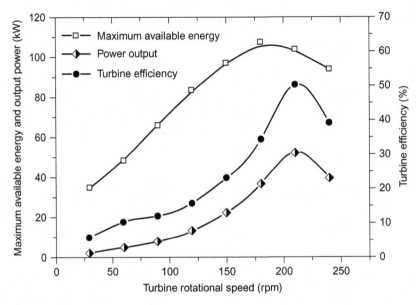

FIGURE 5.7

The effect of rotational speed on the maximum available energy, power output, and turbine efficiency.

we will get higher turbine efficiency and larger power output with the optimization of the turbine blade. Although the 3-blade turbine model is different from the Spanish prototype with 4 blades, there is no doubt that the turbine rotational speed will decrease remarkably when the number of blades of the turbine increases compared with the results shown in Fig. 5.3.

The numerical simulation results shown in Figs. 5.5–5.7, although a little different from the experimental data [2], keep to the principles of heat transfer and flow of free convection system and the operation theory of pressure-staged turbine. Therefore, the numerical simulation method advanced in this chapter is feasible, and the results might be more easily accepted than the 2-D simulation results [14]. In addition, it is a good way to predict the power output of large-scale solar chimney power plant with one or several turbines. The following section shows the results of the MW-graded SCPPSs including a turbine with 5 blades.

5.6.3 RESULTS FOR MW-GRADED SOLAR CHIMNEY

In order to provide a reference for the commercial SCPPSs, numerical simulations have also been carried out for the MW-graded solar chimneys in this section. A physical model of the 5-blade turbine is shown in Fig. 5.2 and the detailed parameters of the CLARK aerofoil are the same as for the kW-graded prototype turbine shown in Tables 5.2 and 5.3. Other dimensions of the MW-graded solar chimney are shown in Table 5.1.

Fig. 5.8 shows the effect of the turbine rotational speed on the average velocity and temperature of the chimney outlet. Similarly, the average velocity of the chimney outlet decreases but the temperature increases with the increase of the rotational speed of the turbine. The main difference between the MW-graded model and the kW-graded model lies in that, with the increase of the rotational speed of the turbine, velocity decreases significantly, while temperature increases slightly. The reasons are as follows: pressure drop across the turbine results in a significant effect on the mass flow rate of the system, and the increase of pressure drop across the turbine can cause a remarkable decrease of the fluid velocity, while the temperature of the chimney outlet changes slightly only because the collector radius is large enough for the air to have enough time to absorb heat energy inside.

Fig. 5.9 shows the effect of the turbine rotational speed on the pressure drop and the mass flow rate across the turbine. As can been seen from this figure, when the turbine rotational speed is lower than 50 rpm, pressure drop across the turbine increases rapidly, while the mass flow rate decreases notably. Inversely, when the rotational speed exceeds 50 rpm, changes of the pressure drop and mass flow rate could be neglected. So it can be estimated, according to the results, that the operation rotational speed of the turbine might be selected as 50 rpm because the power output and turbine efficiency may significantly change with the turbine pressure drop and system mass flow rate across the turbine.

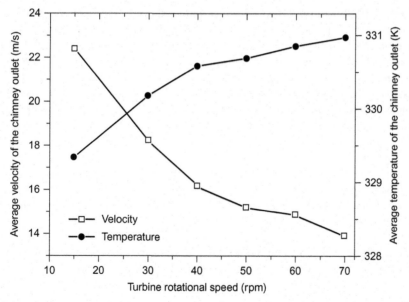

FIGURE 5.8

Outlet parameters of the MW-graded solar chimney power plant.

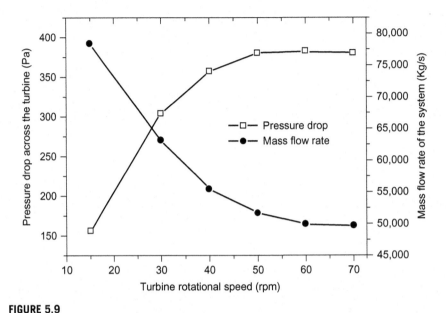

FIGURE 5.9

Pressure drop of the turbine and mass flow rate of the MW graded system.

Fig. 5.10 shows the effect of the turbine rotational speed on the maximum available energy, power output, and turbine efficiency of the MW graded system. It can be easily seen that when the turbine rotational speed is 50 rpm, the maximum available energy reaches the peak value exactly; but the power output and efficiency could not reach their maximum values. When the turbine rotational speed is 40 rpm, the maximum power output surpasses 10 MW, and the maximum turbine efficiency also reaches its peak value, and then the power output and turbine efficiency decrease significantly with the increase of the turbine rotational speed, which means that the best operation rotational speed is 40 rpm for the 5-blade turbine.

Apparently, the 5-blade turbine model designed by the author for MW-grade SCPPS might be unsuitable to be put into practice, as it is difficult to design a single turbine with the output of 10 MW in the present technical conditions. In addition, the aerofoil and detail parameters of the turbine blades need to be further optimized to get much higher power output and turbine efficiency. But there might be a right way in the near future to make a design of several turbines with the rated power output about 6 MW for a large-scale SCPPS.

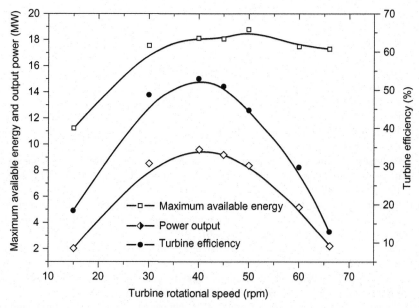

FIGURE 5.10

The effect of rotational speed on the maximum available energy, power output, and turbine efficiency of the MW graded system.

5.7 CONCLUSIONS

A numerical simulation method for the SCPPS including the turbine is presented in this chapter. The results for the Spanish prototype with a 3-blade turbine show that, with the increase of the turbine rotational speed, the average velocity of the chimney outlet and the system mass flow rate decrease, the average temperature of the chimney outlet and the turbine pressure drop inversely, so the maximum available energy, power output, and efficiency of the turbine each have a peak value.

The numerical simulation for the MW-grade SCPPS has been carried out to give a reference for the design of large-scale SCPPSs. For a solar chimney with a chimney 400 m in height and 30 m in radius, a collector 1500 m in radius, and a 5-blade turbine designed in this chapter, the maximum power output and turbine efficiency is about 10 MW and 50%, respectively.

REFERENCES

[1] Haaf W, Friedrich K, Mayer G, Schlaich J. Solar chimneys. Int J Solar Energy 1983; 2:3—20.

[2] Haaf W, Friedrich K, Mayer G, Schlaich J. Solar chimneys. Int J Solar Energy 1984; 2:141—61.

[3] Krisst R. Energy transfer system. Altern Sources Energy 1983;63:8—11.

[4] Kulunk H, A prototype solar convection chimney operated under izmit condition. In: Veziroglu TN, editor. Prod. 7th MICAS, Hemisphere Publishing Corporation, Miami Beach, FL, USA 1985, p. 162.

[5] Pasumarthi N, Sherif SA. Experimental and theoretical performance of a demonstration solar chimney model—part I: mathematical model development. Int J Energ Res 1998;22:277—88.

[6] Pasumarthi N, Sherif SA. Experimental and theoretical performance of a demonstration solar chimney model—part II: experimental and theoretical results and economic analysis. Int J Energ Res 1998;22:443—61.

[7] Padki MM, Sherif SA. On a simple analytical model for solar chimneys. Int J Energ Res 1999;23:345—9.

[8] Lodhi MAK. Application of helio-aero-gravity concept in producing energy and suppressing pollution. Energ Convers Manage 1999;40:407—21.

[9] Bernardes MAD, Valle RM, Cortez MFB. Numerical analysis of natural laminar convection in a radial solar heater. Int J Therm Sci 1999;38:42—50.

[10] Gannon AJ, von Backstrom TW. Solar chimney cycle analysis with system loss and solar collector performance. J Sol Energ-T Asme 2000;122:133—7.

[11] Gannon AJ, von Backstrom TW. Solar chimney turbine performance. J Sol Energ-T Asme 2003;125:101—6.

[12] von Backstrom TW, Gannon AJ. Solar chimney turbine characteristics. Sol Energy 2004;76:235—41.

[13] Bernardes MAD, Voss A, Weinrebe G. Thermal and technical analyses of solar chimneys. Sol Energy 2003;75:511—24.

[14] Pastohr H, Kornadt O, Gurlebeck K. Numerical and analytical calculations of the temperature and flow field in the upwind power plant. Int J Energ Res 2004;28: 495—510.

[15] Schlaich J, Bergermann R, Schiel W, Weinrebe G. Design of commercial solar updraft tower systems—utilization of solar induced convective flows for power generation. J Sol Energ-T Asme 2005;127:117—24.

[16] Ming TZ, Liu W, Xu GL, Yang K. Thermodynamic analysis of solar chimney power plant system. J Huazhong Univ Sci Technolog 2005;33:1—4.

[17] Liu W, Ming TZ, Yang K, Pan Y. Simulation of characteristic of heat transfer and flow for MW-graded solar chimney power plant system. J Huazhong Univ Sci Technolog 2005;33:5—7.

[18] Bilgen E, Rheault J. Solar chimney power plants for high latitudes. Sol Energy 2005; 79:449—58.

[19] Pretorius JP, Kroger DG. Thermoeconomic optimization of a solar chimney power plant. J Sol Energ-T Asme 2008;130.

[20] Ming TZ, Wei L, Xu GL. Analytical and numerical investigation of the solar chimney power plant systems. Int J Energ Res 2006;30:861—73.

[21] Zhou XP, Yang JK, Xiao B, Hou GX. Simulation of a pilot solar chimney thermal power generating equipment. Renew Energ 2007;32:1637−44.

[22] Tao WQ. Numerical heat transfer. 2nd ed. Xi'an, China: Xi'an Jiaotong University Press; 2001.

[23] Launder BE, Spalding DB. The numerical computation of turbulent flows. Comput Methods Appl Mech Eng 1974;3:269−89.

[24] Jayatilleke C. The influence of Prandtl number and surface roughness on the resistance of the laminar sublayer to momentum and heat transfer. Heat Mass Transfer 1969;1:193−321.

Energy storage of solar chimney*

Tingzhen Ming[1,2], Wei Liu[2], Chao Liu[2], Zhou Zhou[2] and Xiangfei Yu[2]

[1]*School of Civil Engineering and Architecture, Wuhan University of Technology, Wuhan, P.R. China* [2]*School of Energy and Power Engineering, Huazhong University of Science and Technology, Wuhan, P.R. China*

CHAPTER OUTLINE

6.1 INTRODUCTION

The solar chimney power plant (SC) system was first proposed in the late 1970s by Professor J. Schlaich and tested with a prototype model in Manzanares, Spain in the early 1980s [1,2]. Compared with the traditional power generation systems, the system has the following advantages: easier to design, more convenient to draw materials, higher operational reliability, fewer running components, more convenient maintenance and overhaul, lower maintenance expense, no environmental contamination, continuous stable running, and longer operational lifespan. It has the potential to meet the power needs of developing countries and territories, especially in deserts where there is a shortage of electric power, with extensive application prospects.

*The content of this chapter was published in Journal of Energy Institute.

Solar Chimney Power Plant Generating Technology. DOI: http://dx.doi.org/10.1016/B978-0-12-805370-6.00006-5

As the SC systems could make significant contributions to the energy supplies of those countries where there is plenty of desert land, which is not being utilized, in recent years, many researchers have made research reports on this technology and have carried out tracking studies on SC systems. Pasumarthi and Sherif [3,4] developed a mathematical model to study the effects of various environment conditions and geometry on the flow and heat transfer characteristics and output power of the solar chimney, and they also developed three different models in Florida and reported the experimental data to assess the viability of the solar chimney concept. Lodhi [5] presented a comprehensive analysis of the chimney effect, power production, efficiency, and estimated the cost of the SC setup in developing nations. Bernardes et al. [6] presented a theoretical analysis of a solar chimney, operating on natural laminar convection under steady state. Gannon and Backström [7] presented an air standard cycle analysis of the SC for the calculation of limiting performance, efficiency, and relationship between the main variables including chimney friction, system, turbine, and exit kinetic energy losses. Gannon and Backström [8] presented an experimental investigation of the performance of a solar chimney turbine. The measured results showed that the solar chimney turbine presented has a total-to-total efficiency of 85−90% and total-to-static efficiency of 77−80% over the design range. Later, the same authors [9] presented analytical equations in terms of turbine flow and load coefficient and degree of reaction, to express the influence of each coefficient on turbine efficiency. Bernardes et al. [10] established a rounded mathematic model for an SC system on the basis of energy-balance principle. Pastohr et al. [11] carried out a two-dimensional steady-state numerical simulation study on the whole SC system which consists of the energy storage layer, the collector, the turbine, and the chimney, and obtained the distributions of velocity, pressure, and temperature inside the collector. Schlaich et al. [12] made an analysis on the operation principle of an SC system and predicted the commercial application prospects of large-scale SC systems. Ming et al. [13] developed a comprehensive model to evaluate the performance of an SC system, in which the effects of various parameters on the relative static pressure, driving force, power output, and efficiency have been further investigated. Pretorius and Kröger [14] evaluated the influence of a developed convective heat transfer equation, more accurate turbine inlet loss coefficient, quality of collector roof glass, and various types of soil on the performance of a large-scale SC system. Bilgen and Rheault [15] designed an SC system for power production at high latitudes and evaluated its performance. Koonsrisuk and Chitsomboon [16] proposed dimensionless variables to guide the experimental study of flow in a small-scale solar chimney and employed a computational fluid dynamics methodology to explore the results that were used to prove the similarity of the proposed dimensionless variables. Maia et al. [17−19] gave detailed theoretical evaluations of the influence of geometric parameters and materials on the behavior of the airflow in a solar chimney prototype and analyzed the airflow characteristics of the systems which can be used as a dryer for agriculture. Ming et al. [20] carried out numerical simulations on the SC systems coupled with a

3-blade turbine using the Spanish prototype as a practical example and presented design and simulation of a MW-graded SC system with a 5-blade turbine, the results of which show that the coupling of the turbine increases the maximum power output of the system and the turbine efficiency is also relatively rather high. Ming et al. [21] established different mathematical models for the collector, the chimney, and the energy storage layer and analyzed the effect of solar radiation on the heat storage characteristics of the energy storage layer. Zhou et al. [22,23] presented some experimental and numerical results of a pilot SC equipment. Ming et al. [24] presented a simple analysis on the thermal performance of solar chimney power generation systems.

The energy storage layer, without which the whole system could not operate continuously during the night, undoubtedly plays a significant role in the power output of a solar chimney system. Part of the solar radiation is absorbed by the energy storage layer during the daytime, and is released during the night or days with cloudy weather. Pastohr et al. [11] presented a numerical simulation result in which the energy storage layer was regarded as solid. The energy storage layer, however, can be treated as aporous medium as there is air flowing inside the solid matrix, especially when the solar chimney systems are built in Gobi or the desert land in the northwest of China with the energy storage layer made of gravel or sand. In this chapter, unsteady conjugate numerical simulations of the solar chimney system with the energy storage layer, the collector, and the chimney are carried out. However, it is not necessary to take the turbine into consideration for a 2-D axisymmetric flow, as the 2-D turbine can only describe the flow and heat transfer characteristics of the system with an ideal pressure drop at a certain place. In addition, the effects of the heat storage characteristics of the energy storage layer on the air flow and heat transfer characteristics of the system is also taken into consideration.

6.2 NUMERICAL MODELS

6.2.1 SYSTEM DESCRIPTION

The physical model of the Spanish solar chimney power generating system prototype shown in Fig. 6.1 [11] is selected as a physical model for the numerical simulation. The prototype has a chimney that is 200 m in height and 5 m in radius, and a collector that is 122 m in radius and 2 m in height. The solid matrix of the energy storage layer is soil or gravel with a large heat capacity.

However, the analysis described in this chapter is based on some simple assumptions shown as follows: (1) Axisymmetric flow of air in the collector is assumed, that is, nonuniform heating of the collector surface in terms of the sun's altitude angle is neglected. (2) An average value for the optical properties is considered to estimate the radiation incident on the absorber surface. Therefore, the transmittance of beam radiation during early sunshine hours would be considerably lower than the average

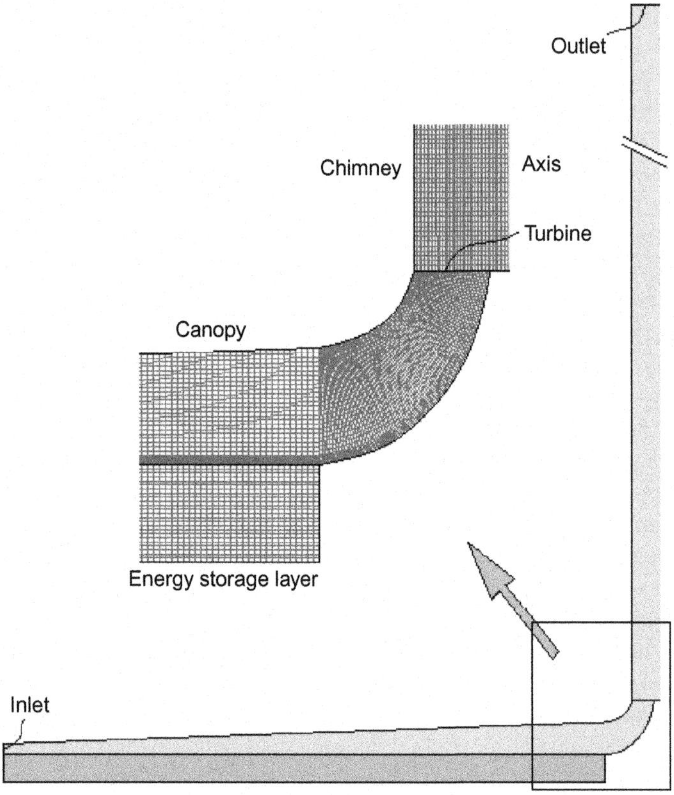

FIGURE 6.1

Physical model of the SC Prototype.

value of the transmittance. (3) The Boussinesq approximation is assumed to be valid. This approximation neglects all variations of properties except for density in the momentum equation. (4) The solar radiation is thought to be transient with time, but the ambient temperature and wind speed are assumed constant. This assumption is advanced only by considering that it is a little difficult to get a convergent simulation result if the ambient parameters were set to be transient with time.

6.2.2 THEORETICAL MODELING

The Continuity equation, Navier-Stokes equation, energy equation, and $\kappa - \varepsilon$ equations can be used to describe the air flow and heat transfer in the collector and chimney shown as follows:

$$\frac{\partial \rho}{\partial t} + \frac{\partial (\rho u)}{\partial x} + \frac{1}{r} \frac{\partial (\rho v)}{\partial r} = 0 \qquad (6.1)$$

$$\frac{\partial(\rho u)}{\partial t} + \frac{\partial(\rho u u)}{\partial x} + \frac{1}{r}\frac{\partial(\rho r v u)}{\partial r} = \rho g \beta(T - T_\infty) + \frac{\partial}{\partial x}\left((\mu + \mu_t)\frac{\partial u}{\partial x}\right) + \frac{1}{r}\frac{\partial}{\partial r}\left((\mu + \mu_t)r\frac{\partial u}{\partial r}\right)$$

$$(6.2)$$

$$\frac{\partial(\rho v)}{\partial t} + \frac{\partial(\rho u v)}{\partial x} + \frac{1}{r}\frac{\partial(\rho r v v)}{\partial r} = \frac{\partial}{\partial x}\left((\mu + \mu_t)\frac{\partial v}{\partial x}\right) + \frac{1}{r}\frac{\partial}{\partial r}\left((\mu + \mu_t)r\frac{\partial v}{\partial r}\right) \qquad (6.3)$$

$$\frac{\partial(\rho T)}{\partial t} + \frac{\partial(\rho u T)}{\partial x} + \frac{1}{r}\frac{\partial(\rho r v T)}{\partial r} = \frac{\partial}{\partial x}\left(\left(\frac{\mu}{Pr} + \frac{\mu_t}{\sigma_T}\right)\frac{\partial T}{\partial x}\right) + \frac{1}{r}\frac{\partial}{\partial r}\left(r\left(\frac{\mu}{Pr} + \frac{\mu_t}{\sigma_T}\right)\frac{\partial T}{\partial r}\right) \qquad (6.4)$$

$$\frac{\partial(\rho \kappa)}{\partial t} + \frac{\partial(\rho \kappa u)}{\partial x} + \frac{1}{r}\frac{\partial(\rho r v \kappa)}{\partial r} = \frac{\partial}{\partial x}\left(\left(\mu + \frac{\mu_t}{\sigma_\kappa}\right)\frac{\partial \kappa}{\partial x}\right) + \frac{1}{r}\frac{\partial}{\partial r}\left(\left(\mu + \frac{\mu_t}{\sigma_\kappa}\right)r\frac{\partial \kappa}{\partial r}\right) + G_\kappa - \rho\varepsilon$$

$$(6.5)$$

$$\frac{\partial(\rho \varepsilon)}{\partial t} + \frac{\partial(\rho \varepsilon u)}{\partial x} + \frac{1}{r}\frac{\partial(\rho r v \varepsilon)}{\partial r} = \frac{\partial}{\partial x}\left(\left(\mu + \frac{\mu_t}{\sigma_\varepsilon}\right)\frac{\partial \varepsilon}{\partial x}\right) + \frac{1}{r}\frac{\partial}{\partial r}\left(\left(\mu + \frac{\mu_t}{\sigma_\varepsilon}\right)r\frac{\partial \varepsilon}{\partial r}\right) + \frac{\varepsilon}{\kappa}(c_1 G_\kappa - c_2\rho\varepsilon)$$

$$(6.6)$$

where, G_κ represents the generation of turbulence kinetic energy due to the mean velocity gradients defined as: $G_\kappa = -\mu_t\left(2\left(\left(\frac{\partial u}{\partial x}\right)^2 + \left(\frac{\partial v}{\partial r}\right)^2 + \left(\frac{v}{r}\right)^2\right) + \left(\frac{\partial u}{\partial r} + \frac{\partial v}{\partial x}\right)^2\right)$. σ_T, σ_κ, and σ_ε are the turbulent Prandtl numbers for T, κ, and ε, respectively, and c_1 and c_2 are two constants for the turbulent model: $c_1 = 1.44$, $c_2 = 1.92$, $\sigma_T = 0.9$, $\sigma_\kappa = 1.0$, $\sigma_\varepsilon = 1.3$, $\mu_t = \frac{c_\mu \rho \kappa^2}{\varepsilon}$, and $c_\mu = 0.09$.

The heat transfer and flow in the energy storage layer may be very complicated, and it is necessary to consider the collector, the chimney, and the storage medium as a whole system. As the material used for energy storage can be regarded as a porous medium, the Brinkman–Forchheimer Extended Darcy model [25] is used to describe the flow in the convective porous-layer, which can be expressed as follows.

$$\frac{\partial\rho}{\partial t} + \frac{\partial(\rho u)}{\partial x} + \frac{1}{r}\frac{\partial(\rho v)}{\partial r} = 0 \qquad (6.7)$$

$$\frac{1}{\varphi}\frac{\partial(\rho u)}{\partial t} + \frac{1}{\varphi^2}\left(\frac{\partial(\rho u u)}{\partial x} + \frac{1}{r}\frac{\partial(\rho r v u)}{\partial r}\right) = \rho g\beta(T - T_e) + \frac{\partial}{\partial x}\left(\mu_m\frac{\partial u}{\partial x}\right)$$
$$+ \frac{1}{r}\frac{\partial}{\partial r}\left(r\mu_m\frac{\partial u}{\partial r}\right) - \frac{\mu u}{K} - \frac{\rho F}{\sqrt{K}}\sqrt{u^2 + v^2}u$$

$$(6.8)$$

$$\frac{1}{\varphi}\frac{\partial(\rho v)}{\partial t} + \frac{1}{\varphi^2}\left(\frac{\partial(\rho u v)}{\partial x} + \frac{1}{r}\frac{\partial(\rho r v v)}{\partial r}\right) = \frac{\partial}{\partial x}\left(\mu_m\frac{\partial v}{\partial x}\right)$$
$$+ \frac{1}{r}\frac{\partial}{\partial r}\left(r\mu_m\frac{\partial v}{\partial r}\right) - \mu_m\frac{v}{r^2} - \frac{\mu u}{K} - \frac{\rho F}{\sqrt{K}}\sqrt{u^2 + v^2}v$$

$$(6.9)$$

$$\rho_m c_{p,m}\left(\frac{\partial T}{\partial t} + \frac{\partial(uT)}{\partial x} + \frac{1}{r}\frac{\partial(rvT)}{\partial r}\right) = \frac{\partial}{\partial z}\left(\lambda_m\frac{\partial T}{\partial x}\right) + \frac{1}{r}\frac{\partial}{\partial r}\left(r\lambda_m\frac{\partial T}{\partial r}\right) \qquad (6.10)$$

where, φ, ρ_m, $c_{p,m}$, μ_m, and λ_m are the porosity, apparent density, specific capacity, dynamic viscosity, and apparent thermal conductivity of the porous medium, respectively: $\rho_m = (1 - \varphi)\rho_s + \varphi\rho_a$, $c_{p,m} = (1 - \varphi)c_{p,s} + \varphi c_{p,a}$, $\lambda_m = (1 - \varphi)\lambda_s + \varphi\lambda_a$, $\mu_m = \mu/\varphi$, the parameters with subscripts s and a denote the corresponding parameters of the solid and air in the energy storage layer, respectively. K, F, and d_b are the permeability, the inertia coefficient, and the particle diameter of the energy storage layer, respectively.

$$K = d_b{}^2\varphi^3/(175(1-\varphi)^2) \tag{6.11}$$

$$F = 1.75\varphi^{-1.5}/\sqrt{175} \tag{6.12}$$

6.2.3 BOUNDARY CONDITIONS AND INITIAL CONDITIONS

(1) Boundary conditions for the side-faces of the energy storage layer

$$\frac{\partial T}{\partial r}\Big|_{r=R} = 0, \quad u = 0, \quad v = 0 \tag{6.13}$$

There might be heat transfer phenomenon between the outside face of the energy storage layer and the material nearby. Therefore, a simplification of this boundary as shown in Eq. (6.13) may overestimate the local temperature profile of the energy storage layer near this location.

(2) Boundary conditions for the chimney wall

$$\frac{\partial T}{\partial r}\Big|_w = 0, \quad u = 0, \quad v = 00 \tag{6.14}$$

(3) Boundary conditions for the collector inlet and chimney outlet

According to the analysis by Pastohr et al. [11], the static pressure difference Δp between the collector inlet and the environment at the same height is 0 Pa, and temperature approximately equals the environment temperature. In addition, the boundary condition for the chimney outlet should be the pressure outlet, and the pressure at this location should also be equal to that of the environment at the same height:

$$\Delta p_{inlet} = 0, \quad T_{inlet} = T_e \tag{6.15}$$

Similar to the method applied by Pastohr et al. [11], the absorption of the solar radiation is considered as a source term in the energy storage layer with a thickness of 0.1 mm. In addition, the boundary condition for the bottom of the energy storage layer could be selected as a constant temperature condition as the temperature distribution 5 m below the surface varies slightly. The boundary conditions for the solar chimney system are shown in Ref. [24].

(4) Initial conditions

It can be easily seen from the boundary conditions shown above that ϕ_{solar} will change with time during a day. Therefore, the solar radiation, the air temperature in the environment, and the initial conditions of the system should be given as follows to analyze the unsteady heat transfer and flow characteristics of the solar chimney system:

$$t = 0, \quad T_e = const, \quad u = 0, \quad v = 0 \tag{6.16}$$

$$\phi_{solar} = \phi_{solar,\max} \sin\left(\frac{t - 1440n}{720}\pi\right) \quad 0 \le t - 1440n < 720 \quad n = 0, 1, \ldots, 4 \tag{6.17}$$

$$\phi_{solar} = 0 \quad 720 \le t - 1440n < 1440 \quad n = 0, 1, \ldots, 4 \tag{6.18}$$

where, n is the ordinal number of the sunny days which varies from 0 to 4, and $\phi_{solar,\max}$ is the maximum value of the solar radiation including the beam radiation, diffuse sky radiation, and ground-reflected radiation.

6.2.4 SOLUTION METHOD

For the present study, the governing Eqs. (6.1)–(6.10) together with all the boundary conditions and initial conditions mentioned above in Eqs. (6.13)–(6.18) were solved with the SIMPLE method by using the commercial software FLUENT 6.3. The standard $k - \varepsilon$ model was used to describe the flow and heat transfer inside the collector and chimney, and the Brinkman–Forchheimer Extended Darcy model was used to describe the flow and heat transfer in the energy storage layer. QUICK format was used for the discretization of the momentum, energy, and other equations. The nonuniform mesh sizes were used for the numerical computation. And for one physical model, we got the grid-independent solution if the grids are about 1.0 million, therefore the grids of the SC in these simulations are about 1.1 million which is acceptable in order to get reliable results.

During the simulation process, $\phi_{solar,\max}$ and T_e are 1000 W/m^2 and 293K, respectively. The temperature of the bottom of the energy storage layer is preset as 300K for a physical model built in Wuhan, one of the three warmest cities in China. The two materials used to analyze the effect of the energy storage layer on the performance of the solar chimney system are soil and gravel. The properties of the soil are as follows: $\rho_{soil} = 1700$ kg/m^3, $c_{p,soil} = 2016$ J/kg K, $\lambda_{soil} = 0.78$ W/m K. The properties of the gravel are: $\rho_{gravel} = 2555$ kgm^3, $c_{p,gravel} = 814.8$ J/kg K, $\lambda_{gravel} = 2.00$ W/m K. The absorptance of the energy storage layer surface is 0.9, and the porosities of the energy storage layers composed of soil and gravel are both selected as 0.3. The particle diameter of the soil and gravel are 0.5 and 4 cm, respectively, and the time step is 5 minutes.

6.3 RELIABILITY OF THE SIMULATION METHOD

In order to confirm the reliability of the numerical simulation carried out in this paper, comparison between the experimental results of the SC prototype in Spain and the simulation results is necessary and carried out as follows. Some parameters

Table 6.1 Comparison Between Measured and Simulated Results

Parameters	Measured	Simulated	Error (%)
Temperature rise of air (K)	25	25.26645	1.0658
Velocity of air at the chimney inlet (m/s)	9	8.812802	−2.079
Output power (kW)	41	40.196	−1.96

of the SC prototype in Spain are set as follows: solar radiation, 1040 W/m^2; absorptance of the ground, 0.56 ~ 0.67; permeability of the collector, 0.8; ambient temperature, 303K. When soil thickness is within 0 ~ 5 cm, the average thermal conductivity is about 0.7 W/m K; when soil thickness is within 5 ~ 10 cm, the average thermal conductivity is about 1.2 W/m K; when soil thickness is within 10 ~ 15 cm, the average thermal conductivity is about 1.5 W/m K.

The comparison between the measured and numerical simulation results is shown in Table 6.1. It can be easily seen that the numerical simulation results agree well with the measured results, which indicates that the simulation method applied in this paper is reliable. The cause for the errors listed in Table 6.1 is the uncertainty of specific parameters such as the density of gasoloid in the air, air humidity, optical parameters, and the property parameters.

6.4 RESULTS AND DISCUSSION

One of the most attractive advantages of the solar chimney systems is the continuous electricity supply regardless of the weather conditions and day–night cycle, and the cardinal part of the system to realize this advantage is the energy storage layer. The thermal performance of the energy storage layer plays an important role in the power output of the solar chimney system with time. It is clear that the power output of the system is affected by the chimney inlet velocity which is nearly equal to the chimney outlet velocity, while the surface temperature of the energy storage layer has significant influence on the air velocity of the chimney inlet, and the average temperature of the whole energy storage layer also notably affects the power output of the system especially when there is no solar radiation at night or on cloudy days. Hence, we should give a detailed description of the temperature distribution of the system and the air velocity of the chimney outlet with the variation of solar radiation.

Fig. 6.2 gives a description of the velocity distributions of the solar chimney system with an energy storage layer made of soil. It can be seen from this figure that the velocity inside the collector and chimney is about several meters per second, whereas it is rather small inside the energy storage layer, and the air velocity of the whole system increases with the solar radiation, and the maximum velocity lies at the bottom of the tower. Simulation results indicate that the temperature inside the

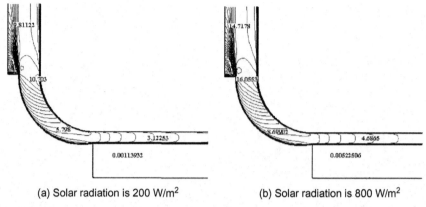

(a) Solar radiation is 200 W/m²   (b) Solar radiation is 800 W/m²

FIGURE 6.2

Velocity distributions of the system under different solar radiation (m/s). (a) Solar radiation is 200 W/m² (b) Solar radiation is 800 W/m².

chimney is only 304K when the solar radiation is 200 W/m², and reaches 321K when the solar radiation is 800 W/m², which shows that solar radiation has a significant influence on the air temperature of the chimney. Further results of numerical simulation with a certain solar radiation and comparison of different models can be found in Ref. [21].

Fig. 6.3 shows the variations of bulk temperature of different energy storage layers with time, where 0 of the abscissa is 06:00 in the morning of the first day. From the figure, we can see that the bulk temperature of the gravel energy storage layer changes more notably than that of the soil energy storage layer during the five days. For a given day, the change scope of bulk temperature of the gravel system is about 1K, which is nearly 80% larger than that of the soil system. This is because the heat capacity of soil is much higher than that of gravel, and therefore the temperature of the gravel energy storage layer changes more notably than that of the soil energy storage layer. As time goes on, the maximum temperature of the soil energy storage layer gradually reaches that of the gravel energy storage layer, and the difference is very small on the fifth day. In contrast, the difference of the minimum temperature between the two types of energy storage layers becomes more and more notable. For the soil energy storage layer, the temperature difference between the day and night decreases gradually, this is because the energy storage increases, and the temperature of the energy storage layer will not decrease significantly even though large amounts of heat energy dissipates from the energy storage layer during the night. In addition, there is a lag effect on the solar radiation for the maximum and minimum temperature of different types of energy storage layer. For the first day, the maximum temperature of the energy storage layer occurs at about 16:00 and the minimum temperature occurs at about 04:00 the next day, but these values come comparatively later in the subsequent

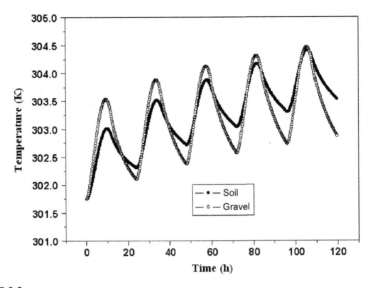

FIGURE 6.3

Variations of bulk temperature of different energy storage layers with time.

days. This phenomenon shows that the energy storage layer has thermal inertia characteristics and has a very notable effect on energy storage. In addition, the overall trend of the bulk temperature seems to increase continuously and show no sign of stabilizing; the reason is that for a 5 m energy storage layer, simulation with 5 days of energy storage is not enough, only when the energy released is equal to the energy storage will the numerical results stabilize.

The variations of the surface temperature of the energy storage layer with time are shown in Fig. 6.4. From this figure, we can see that there is a more notable variation of the surface temperature of the gravel energy storage layer than that of the soil energy storage layer. By comparison, the maximum temperature of the gravel energy storage layer is 15K higher than that of the soil energy storage layer, while the minimum temperature of the former is 2K lower than that of the latter.

The surface temperature of the energy storage layer has a very significant effect on the performance of the solar chimney system. Heat transfer is an irreversible process accompanied with exergy loss to some extent. The higher the surface temperature of the energy storage layer, the larger the temperature difference between the energy storage layer surface and the air inside the collector, and the larger the extent of irreversibility of the heat transfer process, which results in a larger entropy generation. According to the definition of exergy loss, which is the product of environment temperature and entropy generation, the exergy loss caused by the heat transfer process from the higher temperature of the surface of the energy storage layer to the air inside the collector is comparatively larger, and

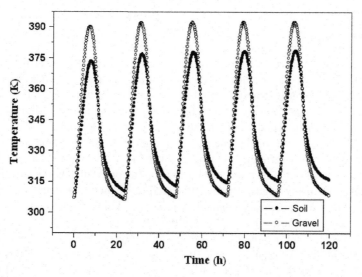

FIGURE 6.4

Variations of the surface temperature of the energy storage layers with time.

the collector efficiency therefore decreases. As a result, shown in Fig. 6.4, the extent of irreversibility caused by the heat transfer process by using the soil energy storage layer is less than that by using the gravel energy storage layer, with a comparatively lower exergy loss. This is very beneficial for the increase of the power output efficiency of the solar chimney system. In addition, the higher the surface temperature of the energy storage layer, if we look inside the temperature gradient of the energy storage layer with a constant temperature of the bottom of the energy storage layer, the larger the temperature difference between the surface and the bottom of the energy storage layer, the greater the temperature gradient inside, and the larger the energy loss from the bottom of the energy storage layer to the deeper soil which could never be used again by the solar chimney system. On the other hand, a comparatively larger temperature gradient inside the energy storage layer will also result in larger exergy loss from the system.

Hence, the effects of the surface temperature of the energy storage layer on the solar chimney systems could be described at least in two aspects: (1) The higher the surface temperature of the energy storage layer, the larger the energy loss from the bottom of the energy storage layer based on the first law of thermodynamics. (2) The higher the surface temperature of the energy storage layer, based on the second law of thermodynamics, the higher the extent of irreversibility during the heat transfer process both inside the energy storage layer and from the energy storage layer surface to the air inside the collector, and the larger the exergy loss during the energy transfer process. Therefore, in order to decrease the energy loss and exergy loss and also to increase the efficiency of the solar chimney system, it is a very effective approach to decrease the surface temperature of the energy storage layer.

There are several methods to decrease the surface temperature of the energy storage layer. In the first way, material with high heat capacity could be selected as the energy storage medium, which could decrease the surface temperature of the energy storage layer effectively. It is thus effective to pave water pipes on the ground inside the collector instead of other energy storage material, since there may be large amounts of energy stored in the water without large temperature differences for the solar chimney system. Secondly, applying material with a comparatively larger thermal conductivity as an energy storage medium can also decrease the surface temperature of the energy storage layer. With the same heat transfer rate, the temperature difference is smaller in the material whose thermal conductivity is higher, and it is therefore useful to adopt a composite energy storage layer with the upper using higher thermal conductivity material and the lower using comparatively lower thermal conductivity, as this kind of energy storage layer could decrease the surface temperature of the energy storage layer with a comparatively smaller amount of energy loss from the bottom to the deeper soil. In addition, different kinds of plants such as flowers and vegetables could be cultivated at different places inside the collector, which can improve the air quality and humidity and also decrease the surface temperature of the energy storage layer.

Figs. 6.5 and 6.6 show the variations of the chimney outlet parameters with time. Air flows inside the collector absorbing energy from the surface of the energy storage layer, and the amount of energy absorbed has an effect on the air temperature and velocity. The larger the amount of energy absorbed by the air, the higher the air temperature and velocity of the airflow of the chimney outlet, while the lower the amount of energy absorbed by the air, the lower the air temperature and velocity of the chimney outlet. When soil is selected as the energy storage medium, the daytime temperature and velocity of the airflow of the chimney outlet from 05:00 to 21:00 will be higher than that when gravel is selected. The temperature and velocity of the outlet at night from 21:00 to 05:00 will be lower for soil than for gravel. The reason is that the thermal conductivity of soil is lower than that of gravel, which results in a comparatively higher temperature difference between the energy storage layer surface and the air inside the collector under the same solar radiation, this creates a comparatively higher heat transfer rate between the energy storage layer surface and the air, where the heat transfer coefficient is corresponding to the air velocity which increases with the increase of the temperature difference between the energy storage layer surface and the air.

The energy storage layer, which can absorb and store solar energy during sunny days and release heat energy to the air inside the collector at nights or on rainy and cloudy days, plays an important role in the continuous fluid flow, heat transfer, and power output of the solar chimney system. Another aim of the energy storage layer is to decrease the system power output difference between the day and night by decreasing the variations of the chimney outlet parameters, especially air velocity. If the variations are too large, the variation of power output of the solar chimney with time will be very large. It can be seen from Figs. 6.5 and 6.6 that the variation scales of the chimney outlet parameters with

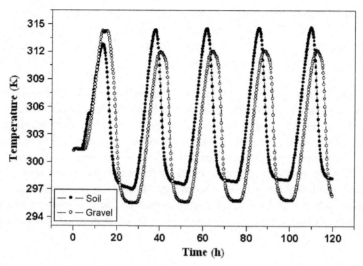

FIGURE 6.5

Variations of the chimney outlet temperature with time.

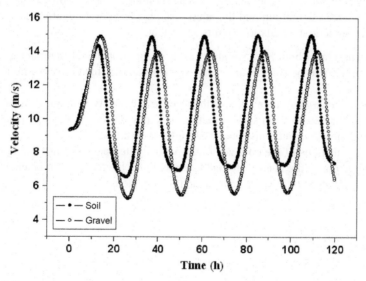

FIGURE 6.6

Variations of the chimney outlet velocity with time.

time are very large. A possible method to decrease these differences is to replace soil and gravel with a kind of energy storage layer made of a new type of composite material whose thermal conductivity decreases as temperature increases, and another possible method is to use a kind of material whose thermal conductivity is higher than that of the gravel but lower than that of the soil. Therefore, heat conduction of the energy storage layer has a complicated effect on the power output characteristic of the solar chimney system. A comparatively higher heat conduction of the energy storage layer will decrease the surface temperature of the energy storage layer, which will result in the decreases of energy loss and exergy loss from the system. On the other hand, a higher heat conduction of the energy storage layer will decrease the heat transfer resistance between the surface of the energy storage layer and the air inside the collector. Hence, care must be taken with the thermal conductivity and heat capacity of the energy storage layer for the power output performance of the solar chimney systems.

Further study will consider the effects of the absorptivity and absorptance of the canopy, and the porosity of the energy storage layer on the performance of the solar chimney system. In addition, research on the performance of the solar chimney system by changing both the turbine pressure drop and the solar radiation will also be valuable.

6.5 CONCLUSIONS

Unsteady numerical simulations on the solar chimney system with an energy storage layer were carried out, and the energy storage layer was regarded as a porous medium. Numerical simulation results show that:

1. Soil and gravel both have suitable values of the property of thermal inertia, and they could be used as an energy storage material for the solar chimney system.
2. When an energy storage layer with larger heat capacity is adopted, a larger part of the energy from solar radiation on sunny days can be stored inside the energy storage layer and released at night or on cloudy days.
3. The fluctuations of the chimney outlet velocity, temperature, and the bulk temperature of the energy storage layer will also decrease when an energy storage layer with a larger heat capacity is adopted.

NOMENCLATURE

A	Collector area, m^2
c	Constant
c_p	Specific heat, kJ/(kg·K)
d_b	Particle diameter of the porous layer, m
F	Inertia coefficient of the porous layer
g	Gravitational acceleration, m/s^2
h	Convection heat transfer coefficient, W/(m^2·K)
K	Permeability of the porous media
p	Pressure, Pa

Pr	Prandtl number
t	Time, s
T	Temperature, K
u	Velocity, m/s
v	Velocity, m/s

GREEK SYMBOLS

α	Thermal diffusivity; absorptance of the collector canopy to the solar radiation energy
β	Thermal expansion coefficient, 1/K
ϕ	Heat transfer rate, W
τ	Transmissivity of the body
σ	Turbulent Prandtl numbers for T, κ, and ε; Stefan-Boltzmann constant.
ρ	Density of the material, kg/m^3
ε	Emissivity of the body surface
φ	Porosity
μ	Dynamic viscosity, kg/(m·s)
λ	Thermal conductivity, W/m
Δ	Difference

SUBSCRIPT

a	Air inside the collector
c	Canopy
chim	Chimney
down	Bottom of the energy storage layer
e	Environment
m	Apparent value of the energy storage layer
s	Solid matrix of the energy storage layer; surface of the energy storage layer
w	Wall of the chimney

REFERENCES

[1] Haaf W, Friedrich K, Mayer G, Schlaich J. Solar chimneys. Int J Solar Energy 1983;2:3–20.

[2] Haaf W, Friedrich K, Mayer G, Schlaich J. Solar chimneys. Int J Solar Energy 1984;2:141–61.

[3] Pasumarthi N, Sherif SA. Experimental and theoretical performance of a demonstration solar chimney model—Part I: mathematical model development. Int J Energ Res 1998;22:277–88.

[4] Pasumarthi N, Sherif SA. Experimental and theoretical performance of a demonstration solar chimney model—Part II: experimental and theoretical results and economic analysis. Int J Energ Res 1998;22:443–61.

[5]   Lodhi MAK. Application of helio-aero-gravity concept in producing energy and suppressing pollution. Energ Convers Manage 1999;40:407−21.

[6]   Bernardes MAD, Valle RM, Cortez MFB. Numerical analysis of natural laminar convection in a radial solar heater. Int J Therm Sci 1999;38:42−50.

[7]   Gannon AJ, von Backstrom TW. Solar chimney cycle analysis with system loss and solar collector performance. J Sol Energ-T Asme 2000;122:133−7.

[8]   Gannon AJ, von Backstrom TW. Solar chimney turbine performance. J Sol Energ-T Asme 2003;125:101−6.

[9]   von Backstrom TW, Gannon AJ. Solar chimney turbine characteristics. Sol Energy 2004;76:235−41.

[10] Bernardes MAD, Voss A, Weinrebe G. Thermal and technical analyses of solar chimneys. Sol Energy 2003;75:511−24.

[11] Pastohr H, Kornadt O, Gurlebeck K. Numerical and analytical calculations of the temperature and flow field in the upwind power plant. Int J Energ Res 2004;28:495−510.

[12] Schlaich J, Bergermann R, Schiel W, Weinrebe G. Design of commercial solar updraft tower systems—utilization of solar induced convective flows for power generation. J Sol Energ-T Asme 2005;127:117−24.

[13] Ming TZ, Wei L, Xu GL. Analytical and numerical investigation of the solar chimney power plant systems. Int J Energ Res 2006;30:861−73.

[14] Pretorius JP, Kroger DG. Critical evaluation of solar chimney power plant performance. Sol Energy 2006;80:535−44.

[15] Bilgen E, Rheault J. Solar chimney power plants for high latitudes. Sol Energy 2005;79:449−58.

[16] Koonsrisuk A, Chitsomboon T. Dynamic similarity in solar chimney modeling. Sol Energy 2007;81:1439−46.

[17] Maia C, Ferreira A, Valle R, Cortez M. Analysis of the airflow in a prototype of a solar chimney dryer. Heat Transfer Eng 2009;30:393−9.

[18] Ferreira AG, Maia CB, Cortez MFB, Valle RM. Technical feasibility assessment of a solar chimney for food drying. Sol Energy 2008;82:198−205.

[19] Maia CB, Ferreira AG, Valle RM, Cortez MFB. Theoretical evaluation of the influence of geometric parameters and materials on the behavior of the airflow in a solar chimney. Comput Fluids 2009;38:625−36.

[20] Ming TZ, Liu W, Xu GL, Xiong YB, Guan XH, Pan Y. Numerical simulation of the solar chimney power plant systems coupled with turbine. Renew Energ 2008;33:897−905.

[21] Zheng Y, Ming TZ, Zhou Z, Yu XF, Wang HY, Pan Y, et al. Unsteady numerical simulation of solar chimney power plant system with energy storage layer. J Energy Inst 2010;83:86−92.

[22] Zhou XP, Yang JK. Temperature field of solar collector and application potential of solar chimney power systems in China. J Energy Inst 2008;81:25−30.

[23] Zhou XP, Yang JK, Xiao B, Long F. Numerical study of solar chimney thermal power system using turbulence model. J Energy Inst 2008;81:86−91.

[24] Ming TZ, Zheng Y, Liu C, Liu W, Pan Y. Simple analysis on thermal performance of solar chimney power generation systems. J Energy Inst 2010;83:6−11.

[25] Kuznetsov AV. Analytical investigation of heat transfer in Couette flow through a porous medium utilizing the Brinkman-Forchheimer-extended Darcy model. ActaMech 1998;129:13−24.

The influence of ambient crosswind on the performance of solar updraft power plant system[1]

Tingzhen Ming[1,2], Xinjiang Wang[2], Jinle Gui[2], Renaud Kiesgen de Richter[3], Wei Liu[2], Guoliang Xu[2], Tianhua Wu[2] and Yuan Pan[4]

[1]*School of Civil Engineering and Architecture, Wuhan University of Technology, Wuhan, P.R. China* [2]*School of Energy and Power Engineering, Huazhong University of Science and Technology, Wuhan, P.R. China* [3]*Institut Charles Gerhardt Montpellier — UMR5253 CNRS-UM2 — ENSCM-UM1—Ecole Nationale Supérieure de Chimie de Montpellier, Montpellier, France* [4]*School of Electrical and Electric Engineering, Huazhong University of Science and Technology, Wuhan, P.R. China*

CHAPTER OUTLINE

[1]The content of this chapter was published in Renewable and Sustainable Energy Reviews and Solar Energy.

Solar Chimney Power Plant Generating Technology. DOI: http://dx.doi.org/10.1016/B978-0-12-805370-6.00007-7

7.1 INTRODUCTION

The solar updraft power plant systems (SUPPS) are among the most sustainable natural resources for electric power generation. They copy the daily solar thermal air motion in the atmosphere to produce electric energy free of CO_2 emissions, and are predicted to be an efficient way of mitigating the unprecedented pressure to reduce CO_2 discharge that many countries in the world are facing today. Compared with more conventional solar energy applications, the SUPPS has gained its ascendancy by achieving several goals: easy to procure building materials, fewer contaminants generated throughout its operating process, and a longer operating lifespan. A SUPPS often consists of four key parts: a collector, a chimney, a turbine, and an energy storage layer. The collector, whose canopy is made of transparent or half-transparent materials, such as plastic and glass, is large enough to collect solar energy because of the one-way-screen characteristics of the canopy material. During the day the visible and UV wavelengths of the solar radiation, where most energy of the sunlight spectrum is concentrated, pass through the cover and warm up the air, meanwhile the infrared wavelengths warm up an energy storage layer. The increased air temperature results in a decrease of its density, while in the meantime a strong airflow buoyant force is produced and the chimney stack leads to natural convection in the SUPP. The air momentum will drive a turbine at the foot of the solar chimney stack whereby kinetic energy will be transformed into electric power. During the night, the heated energy storage layer, which is made of soil, stone, or water in tubes, transfers thermal energy to the air inside the collector, allowing night operation.

The first 50-kW SUPPS prototype built that led to the validation of the solar chimney concept was originally erected in Manzanares, Spain by Jörg Schlaich [1], a professor at the University of Stuttgart, in the early 1980s as a result of a joint venture between the German government and a Spanish utility. The chimney was 194.6 m high and 10 m in radius, with the collector being 122 m in radius and 2−6 m in height from inlet to center. This prototype power station worked successfully for more than 7 years.

Since then, relevant studies on the SUPPS have never ceased due to its prosperous future and some significant breakthroughs in theoretical, numerical analysis and prototype experiments which have been since carried out. All around the world several teams are examining the possibilities of building SUPPS, for instance, Enviromission in Australia and in Arizona (USA) [2,3]. Larbi et al. [4] studied the possible performance of a SUPPS in the southwestern region of Algeria, and Dai et al. [5] in the northwestern regions of China; Zhou et al. [6] did the same for the Qinghai-Tibet region of China; Mostafa et al. [7] estimated the performance of a solar chimney under Egyptian weather conditions; Sangi et al. [8] for Iran, and Ketlogetswe et al. [9] studied the case of Bostwana; Hamdan [10] the Arabian Gulf region; Nizetic [11] looked at the feasibility of implementing SUPPS in the Mediterranean region; and Cervone et al. [12] and Bilgen and Rheault [13] developed a mathematical model to evaluate the performance of SUPPS at high latitudes.

Based on the 50-kW prototype in Manzanares, Spain, Haaf et al. [14,15] made primary investigations into the energy balance, design criteria, and cost analysis in the SUPPS. Following that were the same author's experimental reports on the operating condition of the SUPPS in Spain. In order to analyze the influence of miscellaneous parameters, such as environmental conditions and geometrical dimensions, on the temperature and velocity of air and output power of the solar chimney, the research group led by Professor Sherif [16–21] conducted comprehensive mathematical models to evaluate the fluid flow, heat transfer, and output power performances of various scales of SUPPS and developed three types of experimental prototypes in Florida, with the chimney shape, collector construction, and energy storage layer performance being taken into consideration.

By calculating the performance and efficiency of a SUPPS with chimney friction, turbine, and kinetic energy losses being considered, Gannon and von Backström [22,23] brought forth an air standard cycle analysis of the solar chimney power plant, accompanied by more thorough analysis of SUPPS with turbines being employed by Kröger and Buys [24], Gannon and von Backström [25,26], and Ming et al. [27]. Bernardes et al. [28–30] developed a comprehensive mathematical model to analyze large-scale SUPPS with a double and single collector canopy with an energy storage layer and turbine performance being considered, comparing simulation predictions to experimental results from the prototype plant at Manzanares, and evaluated the operational control strategies applicable to SUPPS. Schlaich et al. [31] presented the basic theory, practical experience, and economy of SUPPS to give a guide for the design of 200-MW commercial SUPPS. Pretorius and Kröger [32,33] developed a comprehensive mathematical model to numerically simulate the SUPPS and analyzed the impact of different calculating methods on output power.

Ming et al. [34] developed a comprehensive model to evaluate the performance of a SUPPS in which the effects of various parameters on the relative static pressure, driving force, power output, and efficiency were further investigated. Zhou et al. [35], Kasaeian et al. [36], and Ferreira et al. [37] conducted various experimental analyses on mini-scale SUPPS. Koonsrisuk and Chitsomboon [38,39] and later Sangi et al. [40] performed detailed theoretical and numerical simulations of SUPPS. Maia et al. [41] theoretically evaluated the influence of some parameters on the behavior of the airflow in a solar chimney indicating that the height and diameter of the chimney are the most important physical variables. Ming et al. [42] and Zheng et al. [43] further studied the power generation and efficiency of the solar chimney power plant systems (SCPPS) coupled with the turbine and discussed effects of different numbers of blades in the Spanish prototype. Also a MW-graded SUPP model was followed by the numerical analysis of the influence of the energy storage layer on the fluid flow, and heat transfer performances of SUPPS by Ming et al. [44] and Xu et al. [45]. As well as this, Ming et al. [46] conducted a numerical analysis on the selection of chimney shape and chimney ratio for a 10 MW SUPPS with the aim of achieving the maximum output power with minimum cost. The results indicated that the cylindrical chimney would be the best choice among the three basic configurations (divergent, conical, and cylindrical chimney), whose optimum H/D value ranges from 6 to 8.

Since 2008, quite a few research studies on SUPPS have been published. Fluri and von Backström [47,48] compared the performance of different turbogenerator layouts, single rotor and counter rotating turbines, both with or without inlet guide vanes, using analytical models and optimization techniques, and discussed the important design parameters. Koonsrisuk and Chitsomboon [39,49−51] predicted the performances of large-scale SUPPSs using the dimensional analysis together with engineering intuition to combine eight primitive variables into only one dimensionless variable that establishes a dynamic similarity between a prototype and its scaled models. Zhou et al. [52] analyzed the influence of chimney height on the performance of SUPPS subjected to standard lapse rate of atmospheric temperature. Zhou et al. [53] and Fluri et al. [54] conducted detailed economic analyses on SUPPSs. Petela [55] conducted a simplified interpretative mathematical model of the SUPPS which could be used to demonstrate feasibility of application of exergy for analysis of SUPPS and for proposing the methodology of the full thermodynamic analysis including exergy. This study presented also the application of the concepts of energy and gravity input for the modified exergetic interpretation of processes. Reports on different applications of conventional SUPPS combined with other systems and sloped SUPPS can be found in Refs. [56−61].

However, a close look at scientific publications shows that researchers focused more on solar effects than on the influence of crosswinds outside the solar chimney. Yet it is generally accepted that, the influence of ambient crosswind on the performance of the SUPPS is self-evident, even as significant as the influence of the solar radiation. Up to now, only a few preliminary studies have been carried out.

Many authors like Niemann and Höffer [62,63] were mainly interested by structural or architectural aspects, concerning mechanics and concrete resistance to wind, vibration, or earthquakes. A model with crosswind was studied by Niemann et al. [64], but was only limited to structural reliability of SUPPS system. Rousseau [65] showed that structural integrity of solar chimneys might be compromised by the occurrence of resonance. The wind gust spectrum peaks near the solar chimney's fundamental resonance frequency which poses a reliability threat, not only to the solar chimney, but also to all high-rise, slender structures. van Zijl and Alberti [66] presented the results of series of physical experiments in wind tunnels establishing external and internal pressure coefficient distributions and overall drag coefficients for rigid smooth cylinders, and demonstrated the stabilizing role of rigid cylinders with vertical ribs. Harte and van Zijl [67] studied the static wind profile converted to pressures acting on the chimney along its height, as well as along the circumference. Lupi [68] and Borri et al. [69] performed innovative modeling of dynamic wind action on SUPPS, and studied structural optimization of solar towers to minimize wind induced effects but again it was limited to structural dynamics aspects. Harte et al. [70] showed that natural draft cooling towers (CTs) and chimneys of SUPPS have many structural properties in common: they are shell structures made of reinforced concrete, they transport warm air by their internal updraft into the atmosphere, and because of their height, gale actions play the most important role in the design and show how far structural design problems of these structures are common. Krätzig et al. [71] explained the design of high efficient tower shells for SUPPS and CT including their critical response characteristics and demonstrated their close structural mechanical relationship to each other. Lv et al. [72] studied SUPPS structure vibrations and frequencies.

Already in the 1970s large dry CTs reaching up to 300 m, were designed for power stations in arid zones. However, at this height they were a long way off the height needed for professional competitive and operational solar chimneys, which start at approximately 700 m.

Although the operation of solar chimneys and CT are quite different, the problems arising from wind gusts and reduction of air-intake-flow-rate under crosswind conditions which decrease the operation efficiency, are well known because of available real scale results obtained from the large number of CT built over the last 40 years. For SUPPS there is little data available, as the tallest prototype 195 m high was built in 1982 in Manzanares (Spain), and since then one smaller one has been built. The promoters of the Arizona solar tower [73], and of the Wuhai Inner Mongolia SUPPS [74] benefit from little and scarce real scale experience and data to strengthen their preliminary studies. The efficiency of SUPPS depends mainly on the size of the collector area and on the height of the chimney, both reasons for their enormous dimensions: collector diameters up to 7 km and chimney heights up to 1500 m are in predesign.

As the concrete cement has to prove its realistic performance by enduring resistance under environmental and operational conditions, constructing CT shells from 100 to 300 m high is a big challenge and the wind load has been well studied. The challenge is enhanced 10-fold for solar updraft towers from 700 to 1500 m high: high compression stresses under dead weight and wind action, suction, or forced wind vibrations in the upper chimney part. The construction of all these very tall chimneys will make high demands on construction techniques and concrete technologies.

But even if CT have greater seniority, until recently the phenomena responsible for wind decreased performance were not well identified and have grabbed the attention of researchers for many years. The way the CT operates under crosswind conditions is still at the forefront of energy research, whereas in the area of SUPPS it is still in its infancy.

For CT, earlier studies made in 1988 by Radosavljevic and Spalding [75], simulated the fluid flow and temperature distribution in a wet CT affected by crosswind. The wind effect on a natural draught dry-CT was studied in 1993 and 1995 by Du Preez and Kröger [76,77], using a simple turbulence model based on the eddy-viscosity to study the flows inside and on the periphery of the tower. In order to minimize the crosswind effect on the thermal performance of towers, they introduced the windbreak walls. Wei et al. [78] in 1995 used complete sample models and wind tunnel testing to demonstrate the adverse effects of wind on the functioning of dry CTs.

Problems encountered in the operation of CTs in the west of Canada led to Derksen et al. [79] investigating in 1996 the effects of wind on the air intake flow rate. Wind tunnel tests on a 1/25 scale model allowed the study of the external flow patterns, pressure characteristics, and air intake flow rate. They discovered that flow imbalance was the cause of ice formation on the inlet side, as the wind-facing side of the tower presented a flow rate almost increased by 45% whereas the opposite rear side flow rate decreased by 19%.

The numerical simulation performed by Su [80] explained the main reason for decline of the performance of dry-CT under crosswind and provided some foundation to improve its thermal performance under crosswind conditions.

Using the commercial software Fluent, Al-Waked and Behnia [81−83], did a 3-D modeling of a tower and simulated the internal and external flows, introducing the wind speed profile as a new important parameter, to correctly simulate the flow around the tower. They also suggested the windbreak walls as a solution for reducing the adverse effects of wind. Zhai and Fu [84,85] performed a numerical and experimental study of the effects of two windbreak sidewalls on the efficiency of a dry-CT. In 2010, Al-Waked [86] investigated the effects of crosswinds on the thermal performance of natural draft wet CTs.

Goodarzi [87−90] demonstrated that a natural draft dry-CT is significantly affected under crosswind conditions and performance might decrease up to 75% in the range of moderate to high wind-velocity conditions. He proposed a new exit configuration for tower stack that could reduce the throttling effect of deflected plume and increase cooling efficiency.

Very little work has been published about crosswind effects on the performance of SUPPS. Serag-Eldin [91] addressed the degradation of SUPPS performance taking into consideration external ambient crosswind, and introduced the concept of controllable flaps aimed to reduce the proportion of hot air gone with the wind by-passing the chimney stack, but did not continue his analysis further on effects of different ambient wind profiles. Pretorius and Kröger [92] analyzed the influence of ambient winds at 2 m/s by regarding its effects on the annual power output of the solar chimney power plant. The comparison of two models which were simulated gave with quiet ambient conditions an annual power output of 373.2 GWh and with windy ambient conditions 336 GWh (a drop of approximately 11%). The windy conditions resulted in an increased convective heat transfer coefficient, facilitating a greater heat flux from the collector roof to the environment and ultimately lower power output.

In this chapter, we numerically analyze the influence of various magnitudes of ambient crosswind on the pressure, velocity, and temperature distributions, and heat transfer and output power performance of SUPPS.

7.2 MODEL DESCRIPTION

7.2.1 GEOMETRIC MODEL

In this paper, a simplified model of the SUPPS Manzanares prototype [14] is adopted for the numerical simulation. As shown in Fig. 7.1, the model has a 200-m-high and 5-m-radial chimney and a collector which covers the earth's surface in a round shape: 120 m in radius and 2 m in height. In order to simulate the performance of the SUPPS exposed to such a vast space, we place the model in the center of an actually nonexistent cubical box with its lengths in x, y, z directions of 400, 400, and 300 m, respectively. A box that has every surface set with different boundary conditions which will be displayed in the ensuing introduction. In this figure, the x-axis is aligned in the velocity direction of the ambient crosswind, and z-axis in the straight up direction. Assuming the symmetric property to be perpendicular to the y-axis direction, only half of the whole system can be taken into consideration as shown in Fig. 7.1, thus the whole geometric dimensions of the cubical box are 400, 200, and 300 m in the x, y, z directions, respectively. Furthermore, in this model, the influence of the energy storage layer was not considered and the geometrical model was not included in the computational model as the present PC could not tolerate the number of meshes of the model including the ambience and the SUPPS with the four parts (collector, turbine, chimney, energy storage layer).

By comparison, the simulation with a much longer X value, say 600 m, downstream of the chimney is also performed, and the influence of this length downstream the chimney on the accuracy of numerical simulation results has been presented in Figs. 7.3 and 7.4. The main aim of this research is to analyze the influence of ambient crosswind on the performance of SUPPS. As for the influence of the outflow from the chimney outlet on the local ambient climate, Zhou et al. [93] has presented some relevant numerical results, and may be one of our further study points will also focus on this.

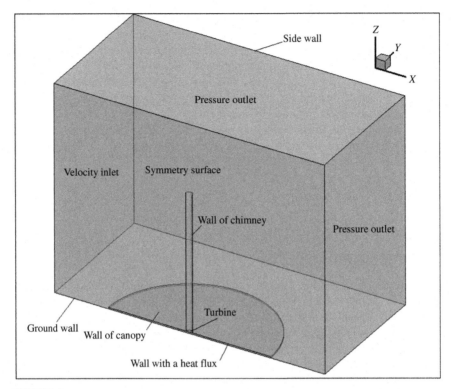

FIGURE 7.1

The 3-D geometrical model including SUPPS and ambience and their boundary conditions.

7.2.2 MATHEMATICAL MODEL

For the SUPPS, we consider the fluid flow inside is natural convection induced by solar radiation heating the ground wall. A criterion number that can measure the intensity of the buoyancy-induced flow is the Rayleigh number being defined as follows:

$$Ra = \frac{g\beta\Delta T L^3}{av} \tag{7.1}$$

where ΔT is the maximum temperature increase within the SUPPS. a, β, and L are the thermal diffusivity, thermal expansion coefficient, and the collector height, respectively. After preliminary calculation, we found that Ra is higher than 10^{10} which is the upper limit transition value of natural convection from laminar to turbulent flow, therefore the turbulent mathematical model needs to be selected to describe fluid flow within the SUPPS. Also, the density variation in the whole computational model including the SUPPS and ambience is so small that it can even be neglected in Mass and Energy equations, making it necessary to resort to Boussinesq approximation without causing large errors [8]. As a result, Mass equation, Navier-Stokes equation, Energy equation, and standard κ-ε equations can be written as follows:

Continuity equation:

$$\frac{\partial \rho}{\partial t} + \frac{\partial (\rho u_i)}{\partial x_i} = 0 \tag{7.2}$$

Navier-Stokes equation:

$$\frac{\partial (\rho u_i)}{\partial t} + \frac{\partial (\rho u_i u_j)}{\partial x_j} = \rho g_i - \frac{\partial p}{\partial x_i} + \frac{\partial \tau_{ij}}{\partial x_j} \tag{7.3}$$

Energy equation:

$$\frac{\partial (\rho c_p T)}{\partial t} + \frac{\partial (\rho c_p u_j T)}{\partial x_j} = \frac{\partial}{\partial x_j}\left(\lambda \frac{\partial T}{\partial x_j}\right) + \tau_{ij}\frac{\partial u_i}{\partial x_j} + \beta T\left(\frac{\partial p}{\partial t} + u_j \frac{\partial p}{\partial x_j}\right) \tag{7.4}$$

Equation for the turbulent kinetic energy k:

$$\frac{\partial}{\partial t}(\rho k) + \frac{\partial}{\partial x_i}(\rho k u_i) = \frac{\partial}{\partial x_j}\left(\alpha_k \mu_{eff}\frac{\partial k}{\partial x_j}\right) + G_k + G_b - \rho \varepsilon - Y_M + S_k \tag{7.5}$$

Equation for the energy dissipation:

$$\frac{\partial}{\partial t}(\rho \varepsilon) + \frac{\partial}{\partial x_i}(\rho \varepsilon u_i) = \frac{\partial}{\partial x_j}\left(\alpha_\varepsilon \mu_{eff}\frac{\partial \varepsilon}{\partial x_j}\right) + C_{1\varepsilon}\frac{\varepsilon}{k}(G_k + C_{3\varepsilon}G_b) - C_{2\varepsilon}\rho\frac{\varepsilon^2}{k} - R_\varepsilon + S_\varepsilon \tag{7.6}$$

where G_k denotes the generation of turbulence kinetic energy because of the mean velocity gradients and can be defined as: $G_k = -\overline{\rho u_i' u_j'}(\partial u_j/\partial x_i)$. σ_T, σ_k, and σ_ε represent the turbulent Prandtl numbers for T, k, and ε, respectively: $\sigma_T = 0.9$, $\sigma_k = 1.0$, $\sigma_\varepsilon = 1.3$. And c_1 and c_2 are two constants for turbulent model: $c_1 = 1.44$, $c_2 = 1.92$. $\mu_t = c_\mu \rho k^2/\varepsilon$ and $c_\mu = 0.09$.

The reason why we select the standard κ-ε equation is that the authors simplified the turbine installed at the base of the chimney as a reverse fan as shown in the following Section 7.2.3. The complex geometrical construction of the three dimensional turbine model within the chimney base will result in complex strong vortexes and turbulent flow which could be accurately simulated using RNG or Realizable κ-ε model [94]. However, in this work, the turbine model is treated as a reverse fan—a two dimensional boundary condition which will not result in complex strong vortexes and turbulent flow—so the standard κ-ε equation is accurate enough to describe the present problem. Detailed numerical simulation coupled with a three dimensional turbine will be performed in a future study.

7.2.3 BOUNDARY CONDITIONS

When taken into account the ambient crosswind on the SUPPS, boundary conditions for both the SUPPS which have been shown in the previous chapters and the ambient should be carefully given. Fig. 7.1 also represents the boundary conditions set in this case as well as the coordinate applied to the model. Detailed descriptions of the boundary conditions are shown as follows.

7.2.3.1 Inlet boundary (surface at x = 0)

The cases in this paper are based on the assumption that the ambient crosswind is fully developed and the temperature constantly at 293K before flowing into the internal space of this model. According to the logarithmic law of the wind speed profile in the atmospheric boundary layer which was proposed by Prandtl in 1932 the ambient wind inlet velocity can be fitted as the equations below [95]:

$$v = w = 0 \tag{7.7}$$

$$u = \frac{1}{\kappa}\left(\frac{\tau_s}{\rho}\right)^{1/2} \cdot \ln\left(\frac{z}{z_0}\right) \tag{7.8}$$

In which τ_s stands for the ground surface shear stress and z_0 for aerodynamic roughness length of the ground. Specific data of z_0 for different terrains is available in charts such as the one that was collected by [95]; and in this case κ and z_0 are chosen as 0.4 and 0.01 m, respectively, since a flat desert terrain type is preset. After that, τ_s can be calculated from a given value of wind speed u at a known height, a height in this case is selected to be that of the tip surface of the SUPPS.

7.2.3.2 Outlet boundary (surfaces at x = 400 and z = 300)

After the influences of the ambience crosswind and the chimney outlet being taken into account, there are two outlet boundaries in the model as shown in Fig. 7.1: the top surface of the model for the exit of the wind from the solar chimney and the surface of the box parallel to that of the inlet ambient crosswind. Pressure outlet boundary condition is applied to each of these and the simplification that reversed flow is normal to the boundary surface.

7.2.3.3 Ground boundary (surface at z = 0)

Both the ground underneath the canopy of the collector and the region outside are contained in the ground boundary condition. The exposed ground outside the collector is assumed to be isothermal and is set to be a wall with a temperature of 318K. This coarse assumption may be more or less influences the accuracy of the simulation results compared to the practical working conditions outside the SUPPS but will not significantly influence the main object of this paper. Consequently, functions for boundary layer are employed here in order to simulate the near-surface flow:

$$u_s = v_s = w_s = 0 \tag{7.9}$$

$$\tau_s = \rho\left[\frac{\kappa}{\ln(z/z_0)}\right]^2 (u^2 + v^2) \tag{7.10}$$

$$k = \frac{\tau_s}{\left(\rho C_\mu^{1/2}\right)} \tag{7.11}$$

$$\varepsilon = \frac{(\tau_s/\rho)^{3/2}}{\kappa y} \tag{7.12}$$

where the subscript s represents "surface value," while other values without such a subscript are evaluated at nodes that are near the ground surface. According to Eqs. (7.11) and (7.12) the values of k and ε are fixed respectively in these zones.

The ground surface covered by the collector canopy is assumed to generate heat fluxes of different values according to various solar radiation intensities. And other parameters set for the wall boundary are quite similar to those of the ground outside the collector. Such simplifications ascribe to the assumptions below:

a. The vibration of wind effects is neglected and thus the time term is considered steady.
b. The solar radiation is uniform vertical incident rays.
c. The ground layer is homogeneous and isotropic.
d. Local heat equilibrium has been achieved between the ground and the air that bypasses it.
e. Radiation heat transfer among the walls of SUPPS in the model is negligible.

All the assumptions introduced are aimed at avoiding analyzing more than one factor simultaneously without deviating too much from the real conditions.

7.2.3.4 Side wall (surface at y = 200)

Due to the long distance of side wall from the solar chimney inlet and outlet, the influence of the side wall is relatively small compared with others listed above. There is hardly any amount of kinetic or thermal turbulence here and thus heat and mass transfer between the geometric model and the outside ambience is weak on the side wall. Accordingly, the parameters of the properties of the side wall are chosen to be default during numerical simulation.

7.2.3.5 Symmetry surface (surface at y = 0)

As mentioned above, because the model in this case is symmetric in the y-direction, and so is the inlet crosswind, symmetry boundary condition is set in this case in order to alleviate the computing process. As is shown in Fig. 7.1, the computing length in the y-direction is shortened to just its half due to the symmetry scheme, thus reducing the computing grids in a large scale. The only necessary procedure to get a whole-zone field is to mirror the simulated velocity and temperature field to the other side of the symmetry surface.

7.2.3.6 Turbine coupling

3-D numerical simulation of the SUPPS coupled with a turbine conducted by Ming et al. [42] indicated that it is a little difficult to simulate the turbine region and many more meshes are needed to accurately describe flow, heat transfer, and output power performances of the system. For the numerical model in this paper it is impossible to realize the simulation procedure simultaneously including regions of the SUPPS, the ambience, and the 3-D turbine due to the limitation of grid numbers. Fortunately, however, the research work conducted by Pastohr et al. [96] indicated that it is also an efficient way to realize the object by simplifying the 3-D turbine to be a 2-D reversed fan with pressure drop across it being preset. This method was verified by Xu et al. [45] and Ming et al. [46] and was proven to be effective to alleviate the mesh pressure by a 3-D turbine region without significantly affecting total performance of SUPPS. Thereby, the turbine is regarded as a reversed fan with pressure drop across it being preset although a 3-D model for the SUPPS and the ambience is selected. To simulate different output power of the turbine, we assign the pressure drop a group of values ranging from 0 to 200 Pa at an interval of 20 Pa. we can thereby calculate the output power of this model according to the equation below:

$$W_e = \eta_t \cdot \Delta p \cdot V \qquad (7.13)$$

where W_e stands for the electric output power of the turbine, η_t represents the overall energy conversion efficiency from thermal to electricity, which is preset as 0.72. This number is the multiple of efficiency from thermal to turbine shaft output power and that from turbine shaft output power to electricity, the former being about 0.8 by Schlaich et al. [31] and the latter being 0.9 which is available easily.

7.2.4 MESHING SKILLS

In general, for the same meshing zone, the hexahedral (HEX) meshing method is more economical and can reduce false diffusion more efficiently than the tetrahedral one. As a result, HEX grids were applied in the model and the mesh generation procedure of the whole geometric model was executed using the commercial software package Gambit 2.3.16. Fig. 7.2 reveals the grid distribution of the geometric model in different angle of view. Fig. 7.2a shows the grid distribution on the symmetric plane. Because of the anticipated relatively steep gradients in velocity, pressure, and temperature in zones such as at the chimney outlet, near the chimney wall and inside the chimney and collector, the grids need to be more concentrated than in the other zones, which is shown in the form as the darker areas. Fig. 7.2b displays the local grid distribution within the chimney and its outside ambience, with finer grids for the boundary layer near the chimney wall, collector wall, and ground wall and the structured quadrilateral grids were adopted in order to reduce the grid number to improve the computing speed without impairing the meshing quality.

(a)   (b)

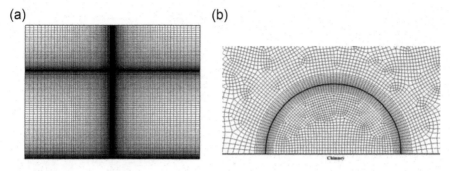

FIGURE 7.2

Grid distribution of the geometric model.

7.2.5 COMPUTATIONAL PROCEDURE

The computations have been performed by using the general purpose CFD program Fluent 6.3.26. The QUICK scheme was used to discretize the convective terms and a second order accurate treatment was used for the diffusion terms. The set of discretized algebraic equations was solved in a coupled manner. The iterations were continued until the relative error in the mass conservation equation was below 5×10^{-5} and in the energy equation 1×10^{-8}. The simulation method using the relative static pressure in place of static pressure to analyze the whole pressure distribution of the whole system is the same as that used by Pastohr et al. [96], Ming et al. [34,42,46], and Sangi et al. [40].

In order to test the grid-independent performance of the grid system selected in the numerical simulation, three test cases of the whole model under the same conditions (ambient crosswind is 0 m/s, and solar radiation is 857 W/m^2) with grid numbers being 1,511,354; 1,674,272; 1,833,458, respectively, were tested. Numerical simulation results indicated that the volume flow rates of the chimney outlet are 789.19, 822.77, and 833.35 m^3/s, respectively, corresponding to the three mesh systems listed above. By comparing the latter two mesh systems, we found that there was only a deviation of approximately 1.3% between these two results, which demonstrated the solutions in this case are grid-independent. The grid spacing and number of 1,674,272 is thus selected as the basic mesh system of this paper.

7.2.6 SELECTION OF AMBIENT GEOMETRICAL DIMENSIONS

The juxtaposed two figures below (Figs. 7.3 and 7.4) reflect the velocity contours in the simulated field. The difference lies in the size of the model, the right one is lengthened in both the x and z directions relative to the left one. Due to the fact that there exists Karman vortex downstream, which would result in the periodic sway of the flow in the rear, it is plausible that the velocity contours behave

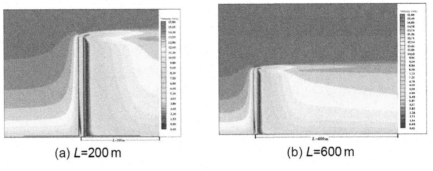

(a) *L*=200 m (b) *L*=600 m

FIGURE 7.3

Influence of ambient geometrical dimensions on system velocity distributions at $G = 857$ W/m^2.

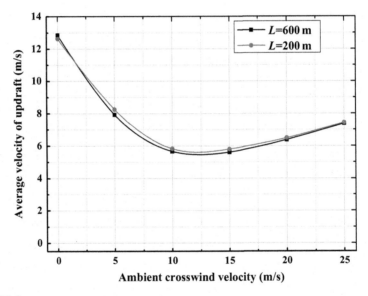

FIGURE 7.4

Influence of ambient geometrical dimensions on chimney outlet velocity at $G = 857$ W/m^2.

unsteadily. Moreover, when the lower left quarter of the larger-scale model is cut off and analyzed, it is found that the contours are identical to those in the small one, thus testifying the reasonability of the results got from a model with just 200 m in the downstream domain.

In order to find the impact of the scale of the simulated domain on the performance of SUPPS, based on the two different model, detailed analysis on the outflow updraft velocity under different ambient conditions, especially velocity, has been carried out. As can be seen from Fig. 7.4, a deviation lower than 3% can be detected from the changing curve of the two models shown in Fig. 7.3. Therefore, the power performance, which is the most important parameter we care about, can be considered as reliable results despite the smaller simulated scale. The appropriate main geometrical dimensions are thus shown in Fig. 7.1.

7.3 RESULTS AND DISCUSSION

For the SUPPS model whose geometrical dimensions mainly come from the prototype in Manzanares, Spain [14], the main factors that influence its characteristics are solar energy input into the system, ambient crosswind, and fore-and-aft pressure drop across the turbine. In this paper, the solar radiation heating the ground surface of the collector can be regarded as heat flux, and wind-velocity magnitude at the top of the chimney and pressure drop across the turbine can be set beforehand.

In this section, the effect of only the ambient crosswind on the performance of SUPPS was researched. A group of wind strength was adopted here: the ambient crosswind velocity at the height of the chimney outlet (200 m) changes from 0 to 15 m/s, at an interval of 5 m/s. According to the logarithmic law of wind profile, in these conditions, the wind velocity at the height of 10 m is 0, 2.2, 4.3, and 6.5 m/s, tantamount to the Beaufort wind force scale 0, 2, 3, and 4, respectively.

Generally, the solar radiation intensity in the deserts of northwest of China varies from 0 to 1400 W/m². When the absorption of the soil and the energy loss being reflected as the form of radiation heat transfer back to space are taken into consideration, the solar energy transferring to the air as thermal energy within the collector is just 60%–70% of the total amount. Therefore, heat flux through the collector bottom ranges from 0 to 980 W/m², making 600 W/m² fairly representative of the common heating condition. Thereby, the heat flux through the ground of the collector was selected as being 600 W/m², namely, the solar radiation is 857 W/m², to analyze the effects of ambient crosswind on the velocity, pressure, and temperature distributions of the system qualitatively.

7.3.1 COMPARISON OF FLOW PERFORMANCES

Fig. 7.5 displays the comparison of contours of velocity magnitude at the symmetry plane in the whole simulated area when the velocity of the ambient crosswind at the height of 200 m increases from 0 to 15 m/s and the solar radiation intensity G remains 857 W/m², constantly, where U_{200m} means the ambient crosswind at the height of 200 m which is also the height of the chimney outlet. It is apparent that when no external wind is blowing horizontally, as shown in Fig. 7.5a, the airflow in the solar chimney and the collector is axisymmetric: air nearby flows into the collector, accelerates gradually and runs centripetally to the foot of the

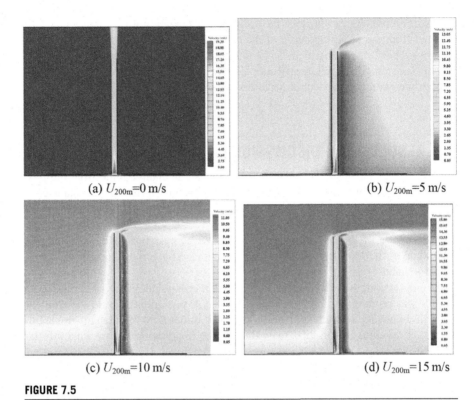

(a) U_{200m}=0 m/s

(b) U_{200m}=5 m/s

(c) U_{200m}=10 m/s

(d) U_{200m}=15 m/s

FIGURE 7.5

Influence of ambient crosswind on velocity distributions in the symmetry plane at $G = 857$ W/m².

chimney with energy being absorbed from the ground surface of the collector and air temperature increasing gradually; then, the heated air is sucked into the chimney and the updraft reaches its peak at a maximum speed of 19.3 m/s at the chimney bottom and finally passes through the chimney smoothly and steadily until it approaches the top outlet of the chimney. However, when the ambient crosswind at the top of the chimney increases to 5 m/s, as shown in Fig. 7.5b, the negative influences are evident: the airflow at the chimney bottom is deflected with a maximum velocity magnitude less than 13.1 m/s; the outflow from the chimney outlet is also slanted downstream by the ambient crosswind instead of rising straightly upward. The same phenomenon occurs when ambient crosswind velocity at the chimney outlet is 10 m/s shown in Fig. 7.5c, namely, the peak air velocity inside the SUPPS decreases to 11.2 m/s with increasing ambient crosswind velocity. However, when approaching 15 m/s, as shown in Fig. 7.5d, the ambient crosswind does not exacerbate the flow performance of SUPPS, instead, it seems to function oppositely: the peak velocity of airflow within the chimney bottom returns to 12.7 m/s, beyond our expectation according to the former downside

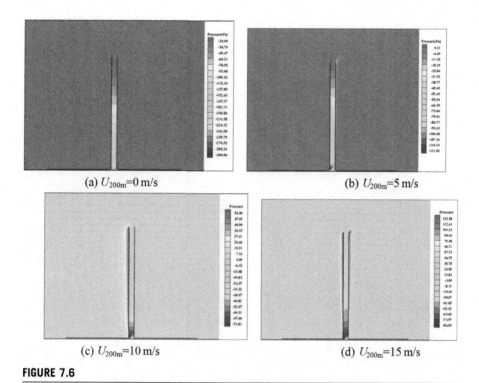

(a) U_{200m}=0 m/s   (b) U_{200m}=5 m/s

(c) U_{200m}=10 m/s   (d) U_{200m}=15 m/s

FIGURE 7.6

Influence of ambient crosswind on pressure distributions in the symmetry plane at $G = 857$ W/m^2.

influence of ambient crosswind. We can thus find that the ambient crosswind has a two-side effect on the SUPPS from the down-and-up performance of airflow velocity magnitude in the chimney.

Fig. 7.6 contracts the static pressure when the ambient velocity differs from each other. From these figures, it is evident that the pressure undergoes a promotion though the chimney, no matter whether there's crosswind outside. Inasmuch as no turbine is installed, wall friction mainly contributes to the depletion of pressure difference as the updraft flows inside.

In order to further illustrate the phenomenon observed above, Fig. 7.7 shows the local velocity vectors near the chimney outlet at $G = 857$ W/m^2 with different ambient crosswind velocities. To some extent, the vectors can reflect the velocity field as their length stands for the velocity magnitude and the arrow heads for the flow direction. As shown in these contrasted figures, the air from the chimney outlet flows straight upwards without ambient crosswind. Then, the outflow slants to the downstream with the existence of ambient crosswind: the higher the ambient crosswind velocity, the more deflective the outflow. It can be expected that, the weak ambient crosswind will inhibit the existent chimney outflow which is

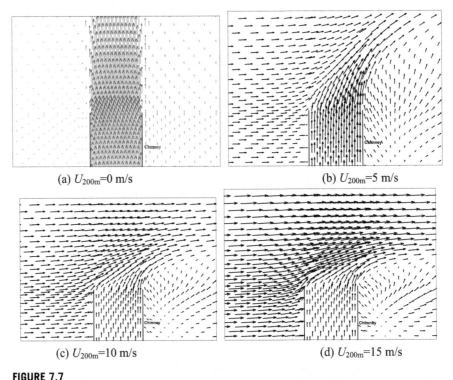

(a) $U_{200m}=0$ m/s (b) $U_{200m}=5$ m/s

(c) $U_{200m}=10$ m/s (d) $U_{200m}=15$ m/s

FIGURE 7.7

Influence of ambient crosswind on local velocity vectors near the chimney outlet at $G = 857$ W/m^2.

due to the buoyancy effect caused by the solar radiation on the collector ground surface, whereas the weak chimney outflow may be strengthened if ambient crosswind is strong enough. A very strong ambient crosswind horizontally flowing across the chimney outlet may produce a negative-pressure zone near the chimney outlet, resulting in an increase of the chimney outlet air velocity.

Fig. 7.8 displays the local velocity vectors at the chimney bottom at $G = 857$ W/m^2 with different ambient crosswind velocities. Comparing Fig. 7.8a to b, we can easily see that airflow within the chimney abates with increasing ambient crosswind, whereas it rises with increasing ambient crosswind from the latter two figures. In addition, there are two phenomena to which we should pay close attention. On the one hand, there is a vortex within the chimney bottom with the existence of the ambient crosswind, and it lies on the left side near the chimney wall as the ambient crosswind flows from the left to right side. This vortex increases its scope with increasing ambient crosswind, leaving the updraft near the chimney wall on the right side being stronger. On the other hand, looking carefully at the airflow direction within the collector, we can find that the airflow

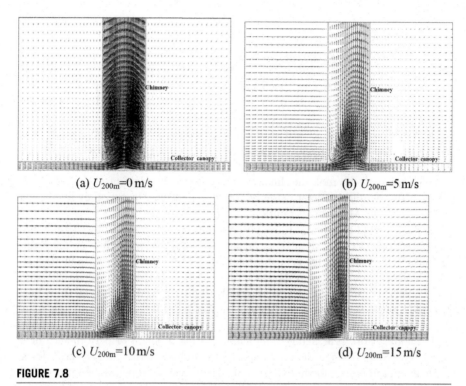

(a) U_{200m}=0 m/s

(b) U_{200m}=5 m/s

(c) U_{200m}=10 m/s

(d) U_{200m}=15 m/s

FIGURE 7.8

Influence of ambient crosswind on local velocity vectors at the chimney bottom at $G = 857$ W/m^2.

sucked from the area in the ambient crosswind's downstream half of the collector, which in the figures is the half on the right, is weakened by the existing ambient crosswind. Especially, when the bottom ambient crosswind is strong enough, air may be blown to the outside of the collector instead of converging into the chimney. Therefore, the enthalpy of the heated air is lost, which might be the main reason why ambient crosswind at low speed has negative effects on the performance of the SUPPS.

Pretorius [97], in an attempt to find a "thermo-economically optimized" dimensional configuration of SUPPS by comparing construction costs to annual power output, tested many plant dimensions with a multitude of variables for different cost structures for SUPPS from 500 to 1500 m high. His numerical model predicted plant configurations for which inflow of cold air into the top of the tower (called cold inflow) may exist in the chimney of the SUPPS, especially when air velocities through the plant are low (very slow upward air velocities). The probability of a plant experiencing cold inflow increases with decreasing collector diameter, chimney height, and collector inlet height, as well as increasing chimney diameter.

In general, for Pretorius' model, if the ratio H/D of the chimney height/chimney diameter is >5 there was no cold inflow, in the absence of crosswind, or with outside wind velocities of 2 m/s at 10 m high and 4 m/s at the top of the chimney. In our model, with a smaller tower (Manzanares model), and higher crosswind velocities from 10 to 20 m/s, inflow of air into the top of the tower seem to be experienced, as seen for the local velocity vectors in Fig. 7.8, at the bottom of the tower.

Serag-Eldin analysis [91] revealed a huge degradation of performance with 10 m/s winds, and even with 2 m/s weak winds a considerable degradation occurs unless the collector inlet height is lower (7.5 m instead of 15 m). As a matter of fact, for his study he took the dimensions suggested by Haaf et al. [14,15] for a 5 MW plant, but the collector entrance is too high (15 m). In our study this height is 2 m like in the Manzanares pilot plant.

In these conditions, Serag-Eldin observed the effect of the atmospheric crosswind is to blow the heated air downstream the collector, rather than up the chimney stack, thus reducing the air temperature in the chimney stack, the temperature difference between hot air in the stack and the atmospheric air outside is lower, the velocity of air in the stack is greatly reduced, and so is the flow rate. Moreover, since this motion is responsible for driving the wind turbine, the plant performance is severely impacted.

In another work [98] Serag-Eldin addressed the problem of performance degradation of SUPPS when they are exposed to strong external atmospheric flows. In an attempt to reduce the hot air escaping sideways he proposed the introduction of controllable flaps (a quarter or a half circle perimeter) at the downstream end of the collector, in order to reduce the proportion of hot air by-passing the chimney stack.

7.3.2 COMPARISON OF RELATIVE STATIC PRESSURE CONTOURS

Fig. 7.9 shows the relative static pressure contours near the chimney bottom at $G = 857$ W/m^2. Because this is the very place where airflow from the collector converges, the pressure undergoes the steepest gradient at the entrance of the chimney. Also, the negative relative static pressure, which represents the pressure difference between the airflow within the chimney and the stable atmosphere outside, has its minimum value at the turning from the collector to the chimney and then increases gradually through the chimney. By analyzing the contrast of pressure distributions as shown from Fig. 7.9a−d, we can find that the minimum relative static pressures at the chimney bottom are −254.37, −118.36, −79.69, and −95.61 Pa, respectively, in corresponding to the four ambient crosswind velocities: 0, 5, 10, and 15 m/s. As the minimum relative static pressure of the SUPPS is a reflection of the system driving force analyzed by Ming et al. [34], the driving force of the SUPPS will also slump at first and then ascends with increasing ambient crosswind velocity. Besides, the reversed flow field of model exposed in strong ambient crosswind is more expansive than in weak ambient

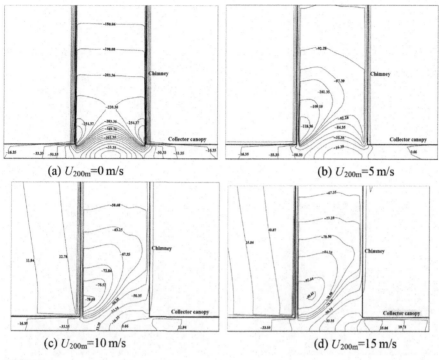

(a) U_{200m}=0 m/s

(b) U_{200m}=5 m/s

(c) U_{200m}=10 m/s

(d) U_{200m}=15 m/s

FIGURE 7.9

Influence of ambient crosswind on local pressure distributions at the chimney bottom at $G = 857$ W/m².

crosswind; in particular, when external crosswind is at a speed of 15 m/s at the top of chimney, the vortex-shedding phenomenon is pretty clear. Also, as shown in Fig. 7.9c,d, the relative static pressure is positive in the downstream half of the collector when the ambient wind is relatively strong, which means that the pressure at this place is higher than that of the ambience at the same height. Thus the air within this zone where the relative static pressure is positive will flow from the collector inlet to the ambience which has been verified previously from the results shown in Fig. 7.8c,d.

7.3.3 COMPARISON OF TEMPERATURE CONTOURS

Fig. 7.10 denotes the contrast of temperature contours in the symmetry plane of SUPPS exposed in different ambient crosswind velocities ranging from 0 to 15 m/s. It is evident that the plume of the air outflow from the chimney outlet deflects more and that its average temperature decreases as the external crosswind becomes stronger: the value of the average temperature is 317.75, 303.84, 299.55, and 297.96 K, respectively, for the four conditions. In addition, the plume scope of the air outflow

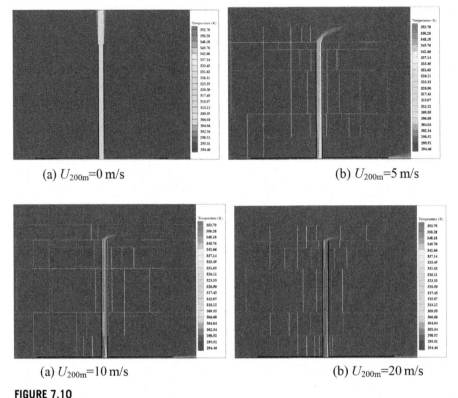

(a) U_{200m}=0 m/s   (b) U_{200m}=5 m/s

(a) U_{200m}=10 m/s   (b) U_{200m}=20 m/s

FIGURE 7.10

Influence of ambient crosswind on temperature distributions in the symmetry plane at $G = 857$ W/m^2.

influenced by the ambient crosswind becomes smaller and smaller with increasing ambient crosswind velocity. This is because: on the one hand, the temperature increase of the updraft decreases with increasing ambient crosswind velocity; on the other hand, convection heat transfer between the plume of the updraft and the ambience has been significantly enhanced with increasing ambient crosswind velocity. From what the comparison displayed, it is safe to draw the conclusion that ambient crosswind has a negative influence on the heat transfer process within the SUPPS, thus reducing the buoyant force originating from the density difference between the SUPPS and the ambience. It is apparent that the difference of temperature in the solar chimney among these conditions partly results from the imposed flow by the ambient crosswind underneath the collector, as shown in Fig. 7.10. The crosswind heading toward the collector outlet in the downstream half of the collector brings about the outflow of heated air. This effect may contribute to the waste of buoyant force generated by heated air and deteriorate the SUPP performance. Furthermore, the external crosswind skating over the chimney wall and collector canopy surface may accelerate the heat transfer between the air within the SUPPS and the ambience, leading to the decrease of the updraft temperature increase, but this effect is comparatively insignificant.

7.3.4 COMPARISON OF SYSTEM TEMPERATURE INCREASE, DRIVING FORCE, AND UPDRAFT VELOCITY

In this section, the numerical simulation results of the geometrical model including SUPPS and ambience in Fig. 7.1 being given different solar intensity and ambient crosswind conditions are quantitatively compared and analyzed as shown from Figs. 7.11−7.13.

Fig. 7.11 displays the influences of external crosswind on the temperature increase of the updraft from the chimney outlet. As seen from this figure, the increase of updraft temperature decreases significantly with the increase of ambient crosswind velocity as long as the solar radiation intensity is over zero, and it increases with increasing solar radiation. Similarly, it is the ambient lowermost crosswind flowing into the collector that gives rise to the decrease of the updraft temperature increase. From the more integral data shown here, it is easy to observe that the SUPPS with higher solar radiation is more sensitive to ambient crosswind: the outlet temperature increase plummets from 30.35K to 13.51K, about 17K drop, when solar radiation is 1143 W/m^2 with ambient crosswind

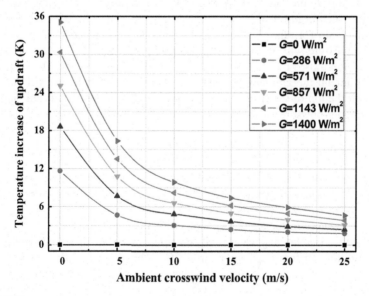

FIGURE 7.11

Influence of ambient wind on the temperature increase of updraft.

velocity at the chimney outlet increasing from 0 to 5 m/s, compared with just 7K drop when the solar radiation is 286 W/m^2. This phenomenon might be mainly caused by increased cooling of the canopy and by the escape of air with relatively high enthalpy from the collector.

Fig. 7.12 shows the influence of the ambient crosswind on the driving force of SUPPS with different solar radiations, where the minimum value of relative static pressure within the chimney bottom can directly reflect the driving force of the SUPPS.

With solar radiation being more than 286 W/m^2 it looks like that with wind speed at 5 and 10 m/s the wind acts as a cover (cap effect) at the tower outlet which decreases the tower airflow.

As shown in this figure, a very interesting phenomenon can be found that, with solar radiation being 0 W/m^2, the driving force of the SUPPS increases gradually with increasing ambient crosswind velocity, whereas it decreases significantly to valley values and then gradually increases with increasing ambient crosswind velocity when the solar radiation is over 0 W/m^2. Apparently the ambient crosswind has positive influence on the driving force of SUPPS with solar radiation being 0 W/m^2, the reason for this might be that the lower the velocity of ambient crosswind that directly enters the collector, the higher the velocity of ambient crosswind that roars past the chimney outlet which will result in a negative-pressure zone near the chimney outlet causing a suction effect of the SUPPS.

The Bernoulli Principle might explain why the higher the velocity of the wind across the top of the chimney of the SUPPS, the faster it will draw air up the chimney (which in turn can increase power output). When wind blows perpendicular to the end of a tube, it creates a vacuum. The higher the wind velocity, the lower the pressure will be, and the more air will be drawn through an open-ended tube. As wind passes over the opening at the top of a chimney, the lower pressure sucks the air up the chimney. In other words with wind blowing with sufficient speed over the top of the chimney, the air pressure at the top is reduced, and tends to draw higher-pressure air from down below. Also the higher the wind speed, the higher the Venturi suction effect at the top of the chimney. The Bernoulli equations are simplifications of the Navier-Stokes Eq. (7.3).

From this figure, when the ambient crosswind velocity is less than 10 m/s, it has negative influence on the driving force of SUPPS; and it will have positive influence on the driving force of SUPPS if it is higher than 10 m/s when the solar radiation is over 0 W/m^2. This is because both negative and positive aspects of ambient crosswind influence the system simultaneously. Apparently, when ambient crosswind is relatively weak, the maximum pressure difference under these solar conditions differs much more from what they do when the ambient crosswind velocity approaches 25 m/s. This phenomenon indicates that with the ambient crosswind getting stronger, solar radiation functions less and, is less crucial in determining the performance of SUPPS while the ambient crosswind becomes more and more predominant.

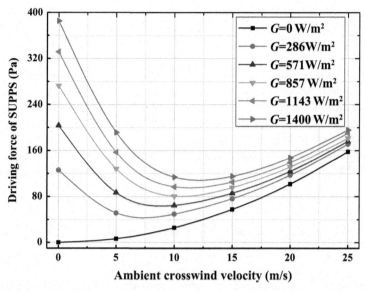

FIGURE 7.12

Influence of ambient wind on the driving force of SUPPS.

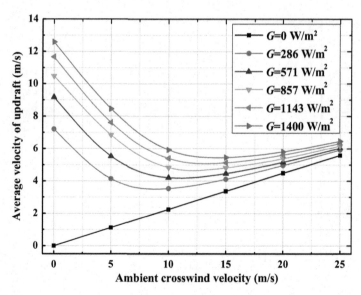

FIGURE 7.13

Influence of ambient crosswind on average velocity of updraft.

Fig. 7.13 displays the relationship between the average velocity of the updraft and the ambient crosswind velocity under different solar radiation intensity. As shown in this figure, the airflow inside the chimney is notably accelerated with increasing ambient crosswind with solar radiation being 0 W/m^2, which is in agreement with the result shown in Fig. 7.12. Similarly, there also exists a valley value of updraft velocity in each line with increasing ambient crosswind when the solar radiation is over 0 W/m^2. Take the curve of G = 1143 W/m^2, for example, we can see that at first the ambient crosswind undermines the average velocity of updraft and further impairs the performance of the SUPPS evidently. However, once the valley point is weathered, the positive effect, namely the suction effect of the SUPPS from the chimney outlet, of external crosswind plays a more important role.

7.3.5 INFLUENCE OF CROSSWIND WITH TURBINE PRESSURE DROP

Fig. 7.14 displays the influence of ambient crosswind on average velocity of updraft with turbine running when the solar radiation is 857 W/m^2, where the basic method for the control of the turbine pressure has been presented by Ming et al. [34] and Xu et al. [45]. As shown in this figure, when the turbine pressure drop remains constant, the relation between average velocity of updraft and the ambient crosswind velocity seems to be similar to that shown in Fig. 7.13. That is, the performance of the solar chimney experiences an overall down-and-up

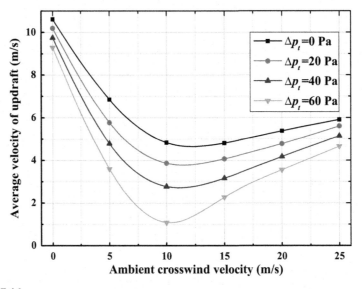

FIGURE 7.14

Influence of ambient crosswind on average velocity of updraft with turbine running at G = 857 W/m^2.

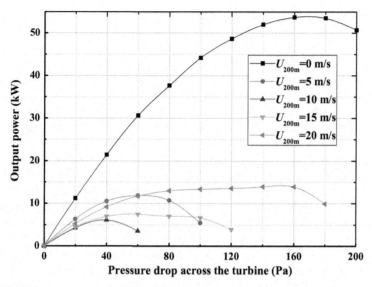

FIGURE 7.15

Influences of pressure drop on turbine output power.

process and the minimum of average velocity of updraft occurs when ambient crosswind velocity is 10 m/s. Numerical simulation results with different solar radiation indicate that the similar phenomenon shown above can be found regardless of the pressure drop, at least in the scope of conditions the model of this paper is involved in. Meanwhile, the average velocity of the updraft decreases notably with increasing turbine pressure drop being given the same ambient crosswind velocity. This phenomenon further denotes that the ambient crosswind and the turbine pressure drop are two key factors that influence the performances of SUPPS independently.

In order to evaluate the output power performance of SUPPS exposed to various ambient crosswind velocities, Fig. 7.15 displays how the output power of the turbine, which represents the ability of the SUPPS to convert the solar and wind energy into electric power, can be influenced by the pressure drop across it at $G = 857$ W/m^2. As for the curve computed in the absence of ambient crosswind, it is rather easy to see that it is so smooth that a parabolic curve of its overall profile can be deduced from the first half of it, similar to what has been observed by Xu et al. [45]. For this phenomenon, a convincing reason is proposed: according to Eq. (7.13), turbine output is determined by both turbine pressure drop and volume flow rate, only when the extent to which pressure drop increases is considerable can the rate of airflow decrease significantly, making the product of them to climb up at first and then run down with the pressure drop.

In other conditions with ambient crosswind velocity being higher than 0 m/s, although the output power curves experience somewhat similarly, they yet differ much in specifics. Differences are the maximum of output power and the corresponding turbine pressure drop, as well as the protruding extent of the curves. When the ambient crosswind velocity is lower than 15 m/s, it has significant influence on the output power: the maximum output power can be over 50 kW at the turbine pressure drop of 160 Pa without ambient crosswind velocity whereas it is lower than 10 kW with ambient crosswind velocity being 10 m/s. It can also be seen from the three output power curves ($U_{200m} = 0$ m/s, $U_{200m} = 5$ m/s, $U_{200m} = 10$ m/s) that their changing trend is very similar. Thus the ambient crosswind has negative influence on the system output power. However, the output power curves of $U_{200m} = 15$ m/s and $U_{200m} = 20$ m/s as shown in Fig. 7.15 differ greatly from the curves with ambient crosswind velocity being lower than 15 m/s. Take the curves of $U_{200m} = 20$ m/s, $U_{200m} = 15$ m/s, and $U_{200m} = 10$ m/s as an example, we found that being given certain turbine pressure drop, the system output power increases with increasing ambient crosswind velocity, which is notably different from the changing trend shown when ambient crosswind velocity is lower than 15 m/s. Thereby, the ambient crosswind has positive influence on the output power of SUPPS. In addition, the maximum output power of SUPPS can be reached at certain turbine pressure drop point with ambient crosswind velocity being lower than 15 m/s, whereas it can be reached during a broad scope of turbine pressure drop with ambient crosswind velocity being higher than 15 m/s. For instance, when the ambient crosswind velocity is 20 m/s, the system output power is about 13−14 kW with the turbine pressure drop changing from 60 to 160 Pa.

7.3.6 MAIN FINDINGS

Through the detailed numerical simulations, this chapter has investigated the influences of various ambient crosswind velocities on the performance of SUPPS. The solar radiation and turbine pressure drop were all being taken into consideration in the analysis. The geometrical model including the SUPPS of the Spanish prototype and its ambience were developed together with the mathematical models describing the fluid flow, heat transfer, and output power performances of the system. The fluid flow, temperature, and pressure distributions as well as the temperature increase, driving force, and output power parameters were also presented and analyzed.

The numerical simulation results revealed that the ambient crosswind has significant influence on the fluid flow, heat transfer, and output power performances of the SUPPS. The feature of the influence has two sides, both positive and negative.

The results have indicated that at low speed the wind effects can be attributed to:

- the thermal flow effect at the level of the greenhouse canopy (wind increases convective heat losses from the collector roof to the environment);
- the entering air at the bottom of the tower causing airflow distortions and resistances, the reduction of pressure or temperature difference between the inside and outside of the tower;
- nonuniform air temperature distribution inside the tower and flow distortions;

- the wind flow covers the exit and deflects the plume. As a result the reduction of the effective area of the chimney outlet causes a throttling or a cap effect which decreases the output air flow;
- possible infiltration of wind into the tower through its top resulting in choking flows inside the tower.

The results seem to indicate that at high speed the wind effects can be attributed to:

- a wind suction effect (Bernoulli type, might be increased by a Venturi deflector) at the tower outlet which increases the tower airflow.

It was found that the SUPPSs performance would be deteriorated if the ambient crosswind velocity was below 15 m/s, and the performance was gradually improved with ambient crosswind velocity being higher than 15 m/s. The trend under different external ambient crosswind conditions may contribute to the optimal design and control for the running of the turbine of SUPPS.

7.4 SC MODEL WITH BLOCKAGE

In this section, the geometrical model was almost similar to that shown in the above. A simplified model of the SUPPS Manzanares prototype [14], was also adopted for the numerical simulation. As shown in Fig. 7.16, the model has a 200-m-high and 5-m-radial chimney and a collector which covers the earth's surface in a round shape: 120 m in radius and 2 m in height. We placed the model in the center of an actually nonexistent cubical box with its lengths in the x, y, z directions of 400, 600, and 300 m, respectively. In this figure, the x-axis was

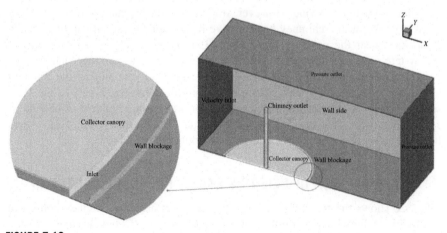

FIGURE 7.16

The 3-D geometrical model including SUPPS and the ambience.

aligned in the velocity direction of the ambient crosswind, the z-axis in the vertical direction. Assuming the symmetric property to be perpendicular to the y-axis direction, only half of the whole system was taken into consideration as shown in Fig. 7.16, thus the whole geometric dimensions of cubical box are 600, 200, and 300 m in the x, y, z directions, respectively. Similarly to the previous research work, the influence of the energy storage layer was not considered.

It should noted that the ambient wind is not fixed, it can enter the SUPPS from the collector inlet in any direction, so the blockage should be circularly built in front of the collector inlet to avoid all the ambient crosswind in any possible directions.

7.5 RESULTS AND DISCUSSION

As mentioned in the previous research work [99], the main factors that influence the characteristics of SUPPS are solar radiation, ambient crosswind, and turbine pressure drop, in which the ambient crosswind significantly affects the key parameters of SUPPS through the collector inlet and chimney outlet. Specially, the contours of temperature, pressure, and velocity within the SUPPS change notably due to the existence of ambient crosswind. In this analysis, the solar radiation is given as 857 W/m^2, the influence of a SUPPS with a blockage wall in front of the collector inlet on the system performance is analyzed.

7.5.1 COMPARISON OF FLOW PERFORMANCES

Figs. 7.17 and 7.18 display the comparison of velocity distributions and pressure distributions at the symmetry plane of the SUPPS with/without blockage. It seems that no significant differences can be seen from the two different systems given the same ambient crosswind. However, from Fig. 7.19, we can see that the velocity vectors within the chimney bottom have been improved significantly by using a blockage near the collector inlet. Fig. 7.19a,c,e shows the velocity vectors of the system without blockage, the outflow near the chimney bottom slants to the downstream with the existence of ambient crosswind: the higher the ambient crosswind velocity, the more deflective the outflow. Further, a part of the fluid flows down to the downstream of the collector with strong ambient crosswind as seen in Fig. 7.19e, whereas this phenomenon has not occurred in Fig. 7.19f where the fluid flows from the collector in all directions to the chimney.

7.5.2 COMPARISON OF RELATIVE STATIC PRESSURE CONTOURS

Fig. 7.20 shows the relative static pressure contours near the chimney bottom of the SUPPS with/without blockage. It seems from Fig. 7.20b,d that the pressure distributions of the system have not been significantly influenced with ambient crosswind being less than 10 m/s. Although the pressure distribution, shown in

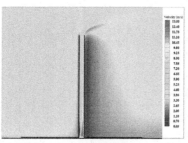 

(a) SUPPS without blockage at U_{200m}=5 m/s   (b) SUPPS without blockage at U_{200m}=15 m/s

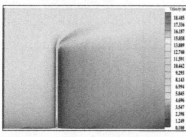 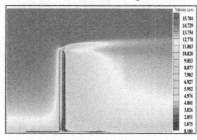

(c) SUPPS with blockage at U_{200m}=5 m/s   (d) SUPPS with blockage at U_{200m}=15 m/s

FIGURE 7.17

Influence of ambient crosswind on velocity distributions in the symmetry plane.

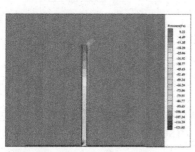

(a) SUPPS without blockage at U_{200m}=5 m/s   (b) SUPPS without blockage at U_{200m}=15 m/s

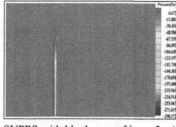 

(c) SUPPS with blockage at U_{200m}=5 m/s   (d) SUPPS with blockage at U_{200m}=15 m/s

FIGURE 7.18

Influence of ambient crosswind on pressure distributions in the symmetry plane.

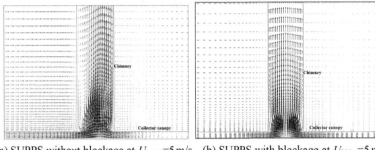

(a) SUPPS without blockage at U_{200m}=5 m/s   (b) SUPPS with blockage at U_{200m}=5 m/s

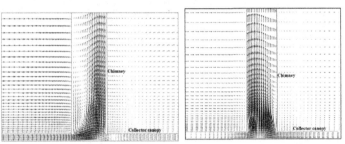

(c) SUPPS without blockage at U_{200m}=10 m/s   (d) SUPPS with blockage at U_{200m}=10 m/s

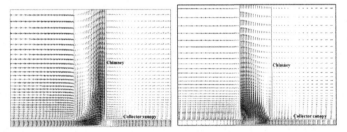

(e) SUPPS without blockage at U_{200m}=15 m/s   (f) SUPPS with blockage at U_{200m}=15 m/s

FIGURE 7.19

Influence of ambient crosswind on local velocity vectors at the chimney bottom.

Fig. 7.20f, is not as good as those shown in Fig. 7.20b,d, which indicates that an ambient crosswind larger than 15 m/s does influence the system pressure distribution, it is still much better than that shown in Fig. 7.20e. Undoubtedly, it is due to the existence of blockage near the collector inlet that all these improvements in pressure distributions of Fig. 7.20b,d,f can be obtained.

7.5.3 FLOW CHARACTERISTICS NEAR THE COLLECTOR INLET

From the comparisons shown in Figs. 7.19 and 7.20 the ambient crosswind entering the system through the collector inlet has an ill effect on the velocity and

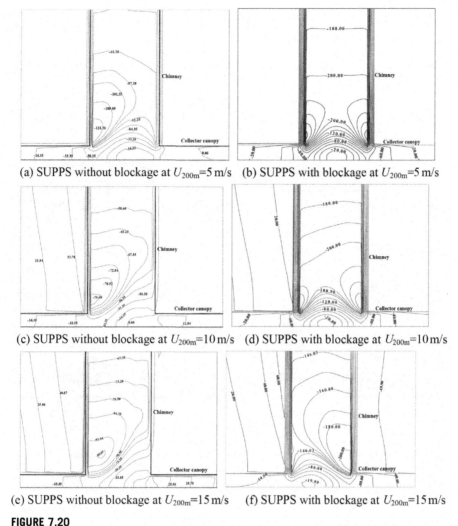

(a) SUPPS without blockage at U_{200m}=5 m/s   (b) SUPPS with blockage at U_{200m}=5 m/s

(c) SUPPS without blockage at U_{200m}=10 m/s   (d) SUPPS with blockage at U_{200m}=10 m/s

(e) SUPPS without blockage at U_{200m}=15 m/s   (f) SUPPS with blockage at U_{200m}=15 m/s

FIGURE 7.20

Influence of ambient crosswind on local pressure distributions at the chimney bottom.

pressure distributions near the chimney bottom, and thereby, it will also has an ill effect on the SUPPS performance. Figs. 7.21−7.23 show the detailed local velocity, temperature, and pressure distributions near the collector inlet of SUPPS with blockage. It is clear that the existence of blockage near the collector inlet significantly changes the local velocity, temperature, and pressure distributions near the collector inlet. The blockage effectively holds back the upstream fluid flow coming from the ambient crosswind and flows outside the collector canopy. Certainly, we can see that a small part of the fluid flow carried by ambient crosswind still enters the collector inlet with ambient crosswind velocity being 15 m/s, and we

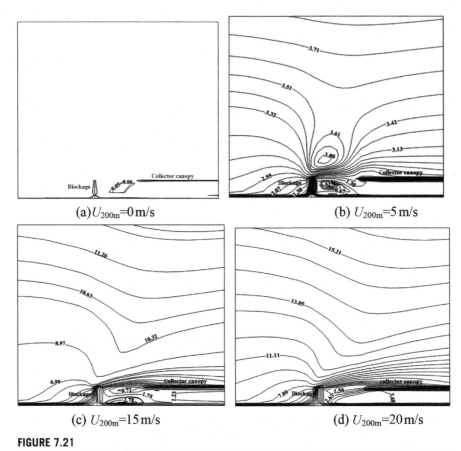

(a) $U_{200m}=0\,\text{m/s}$

(b) $U_{200m}=5\,\text{m/s}$

(c) $U_{200m}=15\,\text{m/s}$

(d) $U_{200m}=20\,\text{m/s}$

FIGURE 7.21

Influence of ambient crosswind on local velocity distributions near the blockage.

can predict that, an increasing percent of fluid flow carried by ambient crosswind will enter the system through the collector inlet with increasing ambient crosswind velocity, which will deteriorate the SUPPS performance.

7.5.4 COMPARISON OF SYSTEM TEMPERATURE INCREASE AND DRIVING FORCE

In this part, numerical simulation results of the two geometrical models, that is, SUPPS with/without blockage and ambience, are quantitatively compared and analyzed as shown in Figs. 7.24 and 7.25.

Fig. 7.24 displays the influence of blockage on the system temperature increase exposed to external ambient crosswind when the turbine is not in operation. As seen from this figure, the increase of updraft temperature decreases

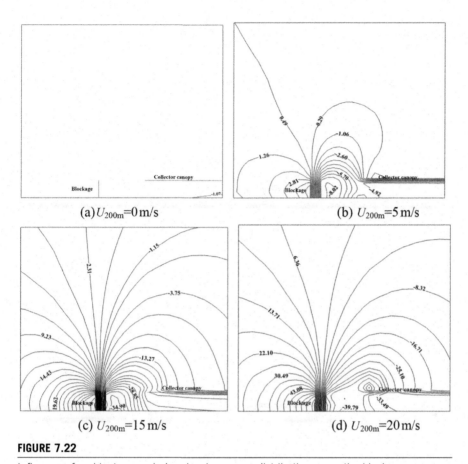

FIGURE 7.22

Influence of ambient crosswind on local pressure distributions near the blockage.

significantly with the increase of ambient crosswind velocity both for these two systems. Whereas, the temperature increase of SUPPS with blockage is much higher than that of SUPPS without blockage at any ambient crosswind velocity: the former is even 10K higher than the latter when the ambient crosswind velocity is 5−15 m/s. This is due to the ambient crosswind velocity entering the system through the collector inlet which significantly changes the fluid flow performance of the SUPPS as shown above. After building a blockage wall a few meters away from the collector inlet, this will to some extent avoid ambient crosswind in any directions influencing the inner pressure, velocity, and temperature distributions of the SUPPS. Similar phenomenon can be seen in Fig. 7.25 which displays the relationship between the driving force of the updraft and the ambient crosswind velocity under given solar radiation intensity. From this figure, we can see that the existence of blockage greatly increases the driving force of the system with ambient crosswind being in the scope of 5−20 m/s.

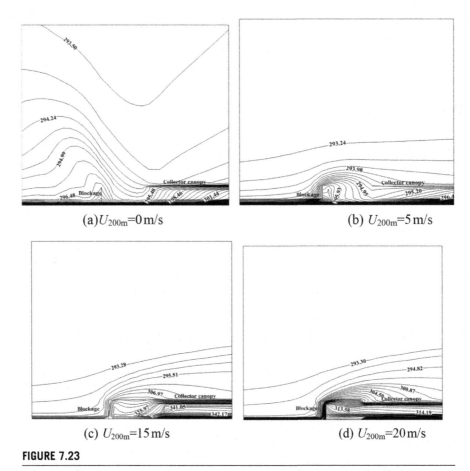

(a)U_{200m}=0 m/s

(b) U_{200m}=5 m/s

(c) U_{200m}=15 m/s

(d) U_{200m}=20 m/s

FIGURE 7.23

Influence of ambient crosswind on local temperature distributions near the blockage.

7.5.5 COMPARISON OF SYSTEM OUTPUT POWER

A detailed comparison of output power with the two different systems has been displayed in Fig. 7.26. It can easily be seen that, when the ambient crosswind velocity is less than 10 m/s, the output powers of the SUPPS with blockage are very close to that of SUPPS without blockage with ambient crosswind velocity being 0 m/s. In other words, at given ambient crosswind velocity higher than 0 m/s, the output power of SUPPS with blockage is much higher than that of SUPPS without blockage, which further verifies that the negative influence of ambient crosswind on the system performance through the collector inlet can be restrained to a great extent.

On the other hand, although the blockage is utilized, the output power of the SUPPS still decreases significantly with increasing ambient crosswind velocity at

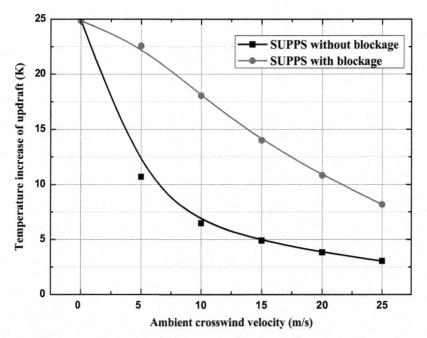

FIGURE 7.24

Influence of ambient wind on the temperature increase of updraft.

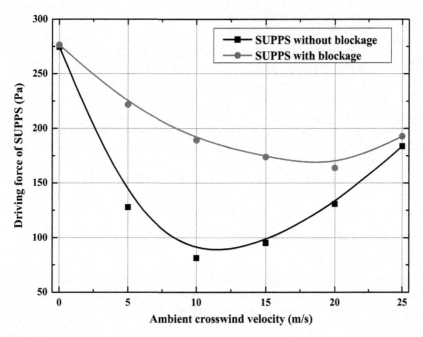

FIGURE 7.25

Influence of ambient wind on the driving force of SUPPS.

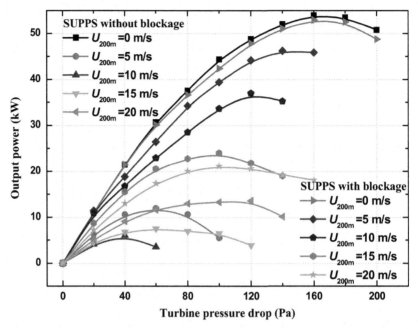

FIGURE 7.26

Influence of pressure drop on turbine output power.

a given turbine pressure drop. This phenomenon is caused by the strong negative influence of ambient crosswind through the chimney outlet. This can be overcome only when the turbine pressure drop is over 160 Pa with ambient crosswind velocity being larger than 20 m/s as shown in Fig. 7.26, but the output power could never reach that without ambient crosswind.

In addition, the comparison shown in Fig. 7.26 further indicates that the ambient crosswind has negative influence on the SUPPS performance both though the collector inlet and though the chimney outlet. The adoption of blockage around the collector inlet can only overcome the negative effect on the SUPPS performance through the collector inlet but the negative effect though the chimney outlet could be overcome by finding other methods.

7.5.6 MAIN FINDINGS

Aiming at decreasing the negative influence of ambient crosswind on the SUPPS performance, a blockage wall was utilized to be built a few meters in front of the collector inlet. The geometrical model including the SUPPS of the Spanish prototype, the blockage and the ambience were developed together with the mathematical models describing the fluid flow, heat transfer, and output power performances of the system. The fluid flow, temperature, and pressure distributions, as well as the temperature increase, driving force, and output power parameters were also presented and analyzed. All the simulation results were compared to the SUPPS without blockage to analyze the effect by utilizing the blockage. The results indicated that the blockage significantly improves the performance of SUPPS exposed to strong ambient crosswind. The velocity, pressure, and temperature distributions within the whole SUPPS have been much better than those of SUPPS without blockage, and therefore, the performance parameters such as temperature increase, driving force, and output power of the SUPPS with blockage all improve significantly at any given ambient crosswind velocity and at any turbine pressure drop.

NOMENCLATURE

$C_{\varepsilon1}, C_{\varepsilon2}, C_{\varepsilon3}$	constants for turbulent model
c_p	specific heat at constant pressure (J/(kg·K))
g	acceleration of gravity (m/s^2)
G	solar radiation intensity (W/m^2)
G_k	turbulence kinetic energy generation due to the mean velocity gradients (J)
G_b	turbulence kinetic energy generation due to turbulence (J)
L	collector height (m)
q	heat flux through the ground underneath the collector (W/m^2)
Ra	Reynolds number (dimensionless)
T	temperature (K)
t	time—for unsteady items (s)
u	velocity in y-direction (m/s)
U_{200m}	ambient crosswind velocity at the height of 200 m (m/s)
w	velocity in z-direction (m/s)
x, y, z	Cartesian space coordinates
a	thermal diffusivity (m^2/s)
υ	kinetic viscosity (m^2/s)
β	volume coefficient of expansion (1/K)
ρ	density (kg/m^3)
τ	shear stress caused by viscosity (N/m^2)
κ	Karman constant

SUBSCRIPTS

s	surface value
i, j	any direction of x, y, and z

REFERENCES

[1] Schlaich J. Solar chimney. Stuttgart: Edition Axel Menges; 1995.

[2] ASX Announcements. <http://www.enviromission.com.au/EVM/content/home.html>; 2005.

[3] E. Limited, Environmission power purchase agreement approved by Southern California public power authority, <http://www.enviromission.com.au/IRM/Company/ShowPage. aspx/PDFs/1154-61988710/MediaReleaseperScppa%20Approval>; 2010.

[4] Larbi S, Bouhdjar A, Chergui T. Performance analysis of a solar chimney power plant in the southwestern region of Algeria. Renew Sust Energy Rev 2010;14: 470−7.

[5] Dai YJ, Huang HB, Wang RZ. Case study of solar chimney power plants in north-western regions of China. Renew Energy 2003;28:1295−304.

[6] Zhou XP, Wang F, Fan JA, Ochieng RM. Performance of solar chimney power plant in Qinghai-Tibet Plateau. Renew Sust Energy Rev 2010;14:2249−55.

[7] Mostafa AA, Sedrak MF, Dayem AMA. Performance of a solar chimney under Egyptian weather conditions: numerical simulation and experimental validation. Energy Sci Technol 2011;1:49−63.

[8] Sangi R. Performance evaluation of solar chimney power plants in Iran. Renew Sust Energy Rev 2012;16:704−10.

[9] Ketlogetswe C, Fiszdon JK, Seabe OO. Solar chimney power generation project—the case for Botswana (Retracted article. See vol. 16, p. 6488, 2012) Renew Sust Energ Rev 2008;12:2005−12.

[10] Hamdan MO. Analysis of a solar chimney power plant in the Arabian Gulf region. Renew Energy 2011;36:2593−8.

[11] Nizetic S, Ninic N, Klarin B. Analysis and feasibility of implementing solar chimney power plants in the Mediterranean region. Energy 2008;33:1680−90.

[12] Cervone A, Zaccagnini Romito D, Santini E. Design of solar chimney power plant for Mediterranean countries. 2011 International Conference on Clean Electrical Power (ICCEP), Ischia, Italy. 2011. p. 14−6.

[13] Bilgen E, Rheault J. Solar chimney power plants for high latitudes. Sol Energy 2005;79:449−58.

[14] Haaf W, Friedrich K, Mayer G, Schlaich J. Solar chimneys. Int J Sol Energy 1983;2:3−20.

[15] Haaf W, Friedrich K, Mayer G, Schlaich J. Solar chimneys. Int J Sol Energy 1984;2:141−61.

[16] Pasumarthi N, Sherif SA. Experimental and theoretical performance of a demonstration solar chimney model—part I: Mathematical model development. Int J Energy Res 1998;22:277−88.

[17] Pasumarthi N, Sherif SA. Experimental and theoretical performance of a demonstration solar chimney model—part II: Experimental and theoretical results and economic analysis. Int J Energy Res 1998;22:443−61.

[18] Padki MM, Sherif SA. On a simple analytical model for solar chimneys. Int J Energ Res 1999;23:345−9.

[19] Padki MM, Sherif SA. Solar chimney for medium-to-large scale power generation. Proceedings of the Manila international symposium on the development and management of energy resources. 1989. p. 432.

[20] Padki MM, Sherif SA. Fluid dynamics of solar chimneys. In: Morrow TB, Marshall LR, Simpson RL, editors. Forum on industrial applications of fluid mechanics, FED-vol. 70. New York: ASME; 1988. p. 43−6.

[21] Padki MM, Sherif SA. A mathematical model for solar chimneys. Proceedings of 1992 international renewable energy conference, Amman, Jordan. 1992. p. 289−94.

[22] von Backström TW, Gannon AJ. Compressible flow through solar power plant chimneys. J Sol Energy-T ASME 2000;122:138−45.

[23] Gannon AJ, von Backström TW. Solar chimney cycle analysis with system loss and solar collector performance. J Sol Energy-T ASME 2000;122:133−7.

[24] Kröger DG, Buys JD. Performance evaluation of a solar chimney power plant. ISES 2001 Solar World Congress, Adelaide, Australia. 2001. p. 907−17.

[25] Gannon AJ, von Backström TW. Controlling and maximizing solar chimney power output. 1st International conference on heat transfer, fluid mechanics and thermodynamics, Kruger Park, South Africa. 2002.

[26] Gannon AJ, von Backström TW. Solar chimney turbine performance. J Sol Energy-T ASME 2003;125:101−6.

[27] Ming TZ, Zheng Y, Liu C, Liu W, Pan Y. Simple analysis on thermal performance of solar chimney power generation systems. J Energy Inst 2010;83:6−11.

[28] Bernardes MAD, von Backström TW, Kröger DG. Analysis of some available heat transfer coefficients applicable to solar chimney power plant collectors. Sol Energy 2009;83:264−75.

[29] Bernardes MAD, von Backström TW. Evaluation of operational control strategies applicable to solar chimney power plants. Sol Energy 2010;84:277−88.

[30] Bernardes MAD, Voss A, Weinrebe G. Thermal and technical analyses of solar chimneys. Sol Energy 2003;75:511−24.

[31] Schlaich J, Bergermann R, Schiel W, Weinrebe G. Design of commercial solar updraft tower systems—utilization of solar induced convective flows for power generation. J Sol Energy-T ASME 2005;127:117−24.

[32] Pretorius JP, Kröger DG. Critical evaluation of solar chimney power plant performance. Sol Energy 2006;80:535−44.

[33] Pretorius JP, Kröger DG. Solar chimney power plant performance. J Sol Energy-T ASME 2006;128:302−11.

[34] Ming TZ, Wei L, Xu GL. Analytical and numerical investigation of the solar chimney power plant systems. Int J Energy Res 2006;30:861−73.

[35] Zhou XP, Yang JK, Xiao B, Hou GX. Experimental study of temperature field in a solar chimney power setup. Appl Therm Eng 2007;27:2044−50.

[36] Kasaeian AB, Heidari E, Vatan SN. Experimental investigation of climatic effects on the efficiency of a solar chimney pilot power plant. Renew Sust Energy Rev 2011;15:5202−6.

[37] Ferreira AG, Maia CB, Cortez MFB, Valle RM. Technical feasibility assessment of a solar chimney for food drying. Sol Energy 2008;82:198−205.

[38] Koonsrisuk A, Chitsomboon T. Dynamic similarity in solar chimney modeling. Sol Energy 2007;81:1439−46.

[39] Koonsrisuk A, Chitsomboon T. Accuracy of theoretical models in the prediction of solar chimney performance. Sol Energy 2009;83:1764−71.

[40] Sangi R, Amidpour M, Hosseinizadeh B. Modeling and numerical simulation of solar chimney power plants. Sol Energy 2011;85:829−38.

[41] Maia CB, Ferreira AG, Valle RM, Cortez MFB. Theoretical evaluation of the influence of geometric parameters and materials on the behavior of the airflow in a solar chimney. Comput Fluids 2009;38:625−36.

[42] Ming TZ, Liu W, Xu GL, Xiong YB, Guan XH, Pan Y. Numerical simulation of the solar chimney power plant systems coupled with turbine. Renew Energy 2008;33:897−905.

[43] Zheng Y, Ming TZ, Zhou Z, Yu XF, Wang HY, Pan Y, et al. Unsteady numerical simulation of solar chimney power plant system with energy storage layer. J Energy Inst 2010;83:86−92.

[44] Ming TZ, Liu W, Pan Y, Xu GL. Numerical analysis of flow and heat transfer characteristics in solar chimney power plants with energy storage layer. Energy Convers Manage 2008;49:2872−9.

[45] Xu GL, Ming TZ, Pan YA, Meng FL, Zhou C. Numerical analysis on the performance of solar chimney power plant system. Energy Convers Manage 2011;52:876−83.

[46] Ming TZ, de Richter RK, Meng FL, Pan Y, Liu W. Chimney shape numerical study for solar chimney power generating systems. Int J Energy Res 2013;37:310−22.

[47] Fluri TP, von Backström TW. Comparison of modelling approaches and layouts for solar chimney turbines. Sol Energy 2008;82:239−46.

[48] Fluri TP, von Backström TW. Performance analysis of the power conversion unit of a solar chimney power plant. Sol Energy 2008;82:999−1008.

[49] Koonsrisuk A, Chitsomboon T. Partial geometric similarity for solar chimney power plant modeling. Sol Energy 2009;83:1611−18.

[50] Koonsrisuk A, Chitsomboon T. A single dimensionless variable for solar chimney power plant modeling. Sol Energy 2009;83:2136−43.

[51] Koonsrisuk A. Comparison of conventional solar chimney power plants and sloped solar chimney power plants using second law analysis. Sol Energy 2013;98:78−84.

[52] Zhou XP, Yang JK, Xiao B, Hou GX, Xing F. Analysis of chimney height for solar chimney power plant. Appl Therm Eng 2009;29:178−85.

[53] Zhou X, Yang HK, Wang F, Xiao B. Economic analysis of power generation from floating solar chimney power plant. Renew Sust Energy Rev 2009;13:736−49.

[54] Fluri TP, Pretorius JP, Van Dyk C, von Backström TW, Kröger DG, van Zijl GPAG. Cost analysis of solar chimney power plants. Sol Energy 2009;83:246−56.

[55] Petela R. Thermodynamic study of a simplified model of the solar chimney power plant. Sol Energy 2009;83:94−107.

[56] Zhou XP, Xiao B, Liu WC, Guo XJ, Yang JK, Fan J. Comparison of classical solar chimney power system and combined solar chimney system for power generation and seawater desalination. Desalination 2010;250:249−56.

[57] Panse SV, Jadhav AS, Gudekar AS, Joshi JB. Inclined solar chimney for power production. Energy Convers Manage 2011;52:3096−102.

[58] Nizetic S, Klarin B. A simplified analytical approach for evaluation of the optimal ratio of pressure drop across the turbine in solar chimney power plants. Appl Energy 2010;87:587−91.

[59] Cao F, Zhao L, Guo LJ. Simulation of a sloped solar chimney power plant in Lanzhou. Energy Convers Manage 2011;52:2360−6.

[60] Chergui T, Larbi S, Bouhdjar A. Thermo-hydrodynamic aspect analysis of flows in solar chimney power plants—a case study. Renew Sust Energy Rev 2010;14:1410−18.

[61] Asnaghi A, Ladjevardi SM. Solar chimney power plant performance in Iran. Renew Sust Energy Rev 2012;16:3383−90.

[62] Harte R, Höffer R, Krätzig WB, Mark P, Niemann HJ. Solar updraft power plants: engineering structures for sustainable energy generation. Eng Struct 2013;56:1698−706.

[63] Niemann HJ, Höffer R. Wind loading for the design of the solar tower. Third international conference on structural engineering, mechanics and computation, Cape Town, South Africa. September 10−12, 2007.

[64] Niemann HJ, Lupi F, Höffer R. The solar updraft power plant: design and optimization of the tower for wind effects. 5th European and African conference on wind engineering, Florence, Italy. July 19−23, 2009.

[65] J.P. Rousseau, Dynamic evaluation of the solar chimney. MScEng-thesis, University of Stellenbosch, South Africa; December 2005.

[66] van Zijl GPAG, Alberti LT. Flow around cylindrical towers: the stabilizing role of vertical ribs. Third international conference on structural engineering, mechanics and computation, Cape Town, South Africa. September 10−12, 2007.

[67] Harte R, van Zijl GPAG. Structural stability of concrete wind turbines and solar chimney towers exposed to dynamic wind action. J Wind Eng Ind Aerod 2007;95:1079−96.

[68] F. Lupi, Structural behaviour, optimization and design of a solar chimney prototype under wind loading and other actions. Master thesis, University of Florence in Cooperation with Ruhr University Bochum; 2009.

[69] Borri C, Lupi F, Niemann HJ. Innovative modelling of dynamic wind action on Solar Updraft Towers at large heights. 8th International conference on structural dynamics, EURODYN 2011, Leuven, Belgium. July 4−6, 2011.

[70] Harte R, Krätzig WB, Niemann HJ. From cooling towers to chimneys of solar upwind power plants. Proceedings of the 2009 Structures Congress, ASCE Conference Proceedings 1−10. 2009. p. 944−53.

[71] Krätzig WB, Harte R, Montag U, Woermann R. From large natural draft cooling tower shells to chimneys of solar upwind power plants. In: Alberto Domingo A, Lazaro C, editors. Proceedings of the International Association for Shell and Spatial Structures (IASS) Symposium. Spain: Universidad Politecnica de Valencia; September 28−October 2, 2009.

[72] Lv XD, Yuan XF, Zhou LA. Wind-induced response of the solar chimney. Adv Struct 2011;163−167(Pts 1−5):4100−3.

[73] R.W. Spence, Update on EnviroMission's Arizona solar tower project. <http://www.drroyspencer.com/2013/06/update-on-enviromissions-arizona-solar-tower-project/>; June 27, 2013.

[74] Chen YS, Yang Y, Wei YL, Yang JH, Tian YR. Solar hot air-flows power generation and its application in Wuhai of Inner Mongolia. Energy Res Info 2010;26:117−22.

[75] Radosavljevic D, Spalding DB. Simultaneous prediction of internal and external aerodynamic and thermal flow field of a natural draft cooling tower in a cross wind. Proceedings of 6th IAHR cooling tower workshop, Pisa. October 1988.

[76] Du Preez AFD, Kröger DG. Effect of wind on performance of a dry-cooling tower. Recovery Syst CHP 1993;13:139−46.

[77] Du Preez AFD, Kröger DG. The effect of the heat exchanger arrangement and windbreak walls on the performance of natural draft dry-cooling towers subjected to cross-winds. J Wind Eng Ind Aerod 1995;58:293−303.

[78] Wei QD, Zhang BY, Liu KQ, Du XD, Meng XZ. A study of the unfavorable effects of wind on the cooling efficiency of dry cooling towers. J Wind Eng Ind Aerod 1995;54:633−43.

[79] Derksen DD, Bender TJ, Bergstrom DJ, Rezkallah KS. A study on the effects of wind on the air intake flow rate of a cooling tower: Part 1. Wind tunnel study. J Wind Eng Ind Aerod 1996;64:47−59.

[80] Su MD, Tang GF, Fu S. Numerical simulation of fluid flow and thermal performance of a dry-cooling tower under cross wind condition. J Wind Eng Ind Aerodyn 1999;79:289−306.

[81] Al-Waked R, Behnia M. The performance of natural draft dry cooling towers under crosswind: CFD study. Int J Energ Res 2004;28:147−61.

[82] Al-Waked R, Behnia M. CFD simulation of wet cooling towers. Appl Therm Eng 2006;26:382−95.

[83] Al-Waked R, Behnia M. The effect of windbreak walls on the thermal performance of natural draft dry cooling towers. Heat Transfer Eng 2005;26:50−62.

[84] Zhai Z, Fu S. Improving cooling efficiency of dry-cooling towers under cross-wind conditions by using wind-break methods. Appl Therm Eng 2006;26:1008−17.

[85] Fu S, Zhai ZQ. Numerical investigation of the adverse effect of wind on the heat transfer performance of two natural draft cooling towers in tandem arrangement. Acta Mech Sin 2001;17:24−34.

[86] Al-Waked R. Crosswinds effect on the performance of natural draft wet cooling towers. Int J Therm Sci 2010;49:218−24.

[87] Goodarzi M, Ramezanpour R. Alternative geometry for cylindrical natural draft cooling tower with higher cooling efficiency under crosswind condition. Energy Convers Manage 2014;77:243−9.

[88] Goodarzi M, Keimanesh R. Heat rejection enhancement in natural draft cooling tower using radiator-type windbreakers. Energy Convers Manage 2013;71:120−5.

[89] Goodarzi M, Amooie H. A proposed heterogeneous distribution of water for natural draft dry cooling tower to improve cooling efficiency under crosswind. 2012 4th Conference on Thermal Power Plants (CTPP). 2012.

[90] Goodarzi M. A proposed stack configuration for dry cooling tower to improve cooling efficiency under crosswind. J Wind Eng Ind Aerod 2010;98:858−63.

[91] Serag-Eldin MA. Mitigating adverse wind effects on flow in solar chimney plants. ASME heat transfer/fluids engineering summer conference, Charlotte, NC. July, 2004.

[92] Pretorius JP, Kröger DG. The influence of environment on solar chimney power plant performance. Res Develop J S Afr Inst Mech Eng 2009;25:1−9.

[93] Zhou XP, Yang JK, Ochieng RM, Li XM, Mao B. Numerical investigation of a plume from a power generating solar chimney in an atmospheric cross flow. Atmos Res 2009;91:26−35.

[94] Fluent, Inc., FLUENT 6.3 user's guide. <https://www.sharcnet.ca/Software/Fluent6/html/ug/main_pre.htm>; 2006.

[95] Cermak JE. Applications of fluid mechanics to wind engineering-freeman scholar lecture. ASME J Fluids Eng 1975;97:9−38.

[96] Pastohr H, Kornadt O, Gurlebeck K. Numerical and analytical calculations of the temperature and flow field in the upwind power plant. Int J Energy Res 2004;28:495−510.

[97] J.P. Pretorius, Optimization and control of a large-scale solar chimney power plant. Ph.D. thesis, University of Stellenbosch, South Africa, 2007. 212.

[98] Serag-Eldin MA. Mitigating adverse wind effects on flow in solar chimney plants. Proceedings of 4th IEC, Mansoura international engineering conference, Sharm ElSheikh. April 20–22, 2004.

[99] Ming TZ, Wang XJ, de Richter RK, Liu W, Wu TH, Pan Y. Numerical analysis on the influence of ambient crosswind on the performance of solar updraft power plant system. Renew Sust Energy Rev 2012;16:5567–83.

Experimental investigation of a solar chimney prototype

Tingzhen Ming[1,2], Wei Liu[2], Yuan Pan[3] and Zhou Zhou[2]

[1]*School of Civil Engineering and Architecture, Wuhan University of Technology, Wuhan, P.R. China* [2]*School of Energy and Power Engineering, Huazhong University of Science and Technology, Wuhan, P.R. China* [3]*School of Electrical and Electric Engineering, Huazhong University of Science and Technology, Wuhan, P.R. China*

CHAPTER OUTLINE

8.1 INTRODUCTION

The solar chimney power generation system (SC), which has the following advantages while compared with the traditional power generation systems: easier to design, more convenient to draw materials, lower cost of power generation, higher operational reliability, fewer running components, more convenient maintenance and overhaul, lower maintenance expense, no environmental contamination, continuous stable running, and longer operational lifespan, is a late-model solar power generation system [1].

No related experimental results on large-scale commercial SC system, however, have ever been reported since the first SC prototype was built in Spanish in 1980s [2,3], which is mainly because of the excessive early cost required. Establishing a large-scale commercial SC of about 200 MW output power requires the financial support from both local government and enterprise. In 1985, Kulunk [4] set up a

Solar Chimney Power Plant Generating Technology. DOI: http://dx.doi.org/10.1016/B978-0-12-805370-6.00008-9

miniature SC experimental facility. In 1997, Sherif et al. [5−6] set up three SC models by modifying the shape and radius of the collector or canopy in Gainesville of Florida University, and carried out experiments on the temperature and velocity distributions of the airflow inside the canopy, whose results agree well with the theoretical analysis. Zhou et al. [7] presented a comparison of experimental results and simulation results of a pilot SC equipment.

As for the study of theoretical and numerical analysis of the SC systems, many researchers have conducted related mathematical models and simulation results on different kinds of SC systems. Bernardes et al. [8] established a rounded mathematic model for SC system on the basis of the energy-balance principle. Pastohr et al. [9] carried out a two-dimensional steady-state numerical simulation study on the whole SC system which consists of the energy storage layer, the collector, the turbine, and the chimney, and obtained the distributions of velocity, pressure, and temperature inside the collector. Denantes et al. [10] developed and validated an efficiency model at design performance for counter-rotating turbines. Ming et al. [11] developed a comprehensive model to evaluate the performance of an SC system, in which the effects of various parameters on the relative static pressure, driving force, power output, and efficiency have been further investigated. Ming et al. [12,13] established different mathematical models for the collector, the chimney, and the energy storage layer and analyzed the effect of solar radiation on the heat storage characteristic of the energy storage layer. Ming et al. [14] carried out numerical simulations on the SC systems coupled with a three-blade turbine using the Spanish prototype as a practical example and presented the design and simulation of a MW-graded SC system with a five-blade turbine, the results of which show that the coupling of the turbine increases the maximum power output of the system and the turbine efficiency is also relatively rather high. Later, the same authors [15] presented a simple analysis on the thermodynamics of the solar chimney systems.

In this chapter an indoor miniature SC experimental facility is introduced. There are two primary differences between this facility and other experimental facilities shown above: One is that several heaters, replacing the effect of solar radiation, are used to supply energy for the air inside the collector, and the other is that the canopy of this facility is made of heat adiabatic material instead of transparent material as it is unnecessary to receive any energy from the outside. The main work done in this paper includes measurement of temperature and velocity distributions of airflow inside the SC prototype, and analysis of the effects of heat flux exerted on the collector, time and the ambient temperature on the air flow, and heat transfer characteristics of this prototype model.

8.2 EXPERIMENTAL SETUP

Fig. 8.1 shows the experimental setup for a miniature SC prototype model. The system is made up of collector canopy, chimney, heaters, and a ground thermal insulating layer. The chimney, of height 2500 mm, outside diameter 106 mm, and

FIGURE 8.1

Experimental setup of the solar chimney system.

wall 2 mm thick, is made of polytetrafluoroethylene; the collector canopy is made of foam which is 30 mm thick and a square 2000 mm in length, the periphery of the canopy is open to the environment with a height of 55 mm; two pieces of rectangular iron sheet of length 2000 mm and width 1000 mm are applied at the bottom of the collector for better heat transfer, under which lies a closed heating space with eight plate electric heaters uniformly installed inside, the resistance of each heater is about 60 Ω, the material under the bottom of the closed space is an adiabatic layer made of perlite with a thickness of 55 mm.

Related laboratory instruments during experiment are shown as follows: 16 copper constantan thermocouples and a 5.5 digit-voltmeter are applied for the temperature measurement of the air inside the collector, chimney, and the environment; a hot wire anemoscope of Testo Company is applied for the measurement of ambient wind and the air velocity within the chimney; and the solar radiation is simulated through setting the voltage of an autocouple voltage regulator.

8.3 DISPOSAL OF MEASUREMENT POINTS

As for a SC prototype, the measurement of the system temperature and velocity distributions becomes the most important experimental target. Temperature

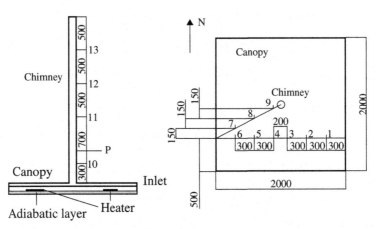

FIGURE 8.2

Dispersal of measurement points of the system (mm). (a) Front view and (b) aerial view.

measurement points were dispersed as shown in Fig. 8.2, among which nine were dispersed inside the collector, four were along the chimney. Moreover, one measurement point was placed on the surface of the collector canopy, one was on the outside wall of the chimney, and one was for the ambient temperature. There were 16 temperature measurement points used to detect the temperature variation of the SC system and the ambient.

For abscissa, from east to west; for vertical ordinate, from south to north, locations for temperature measurement points were shown as follows (units for all measurement points adopt mm): 1(300, 500), 2(600, 500), 3(900, 500), 4(1100, 500), 5(1400, 500), 6(1700, 500), 7(1700, 650), 8(1400, 800), 9(1100, 950). Locations for measurement points along the chimney, from the bottom up, were as follows: 10(1000, 300), 11(1000, 1000), 12(1000, 1500), 13(1000, 2000). Measurement point 14 was on the outer wall of the chimney, measurement point 15 on the outer wall of the collector canopy, and point 16 was used to measure the ambient temperature. There was a little pore used as a chimney air velocity measurement point: P (1000, 500).

8.4 RESULTS AND DISCUSSION

8.4.1 VARIATIONS OF TEMPERATURE WITH TIME

Measurement conditions for Fig. 8.3 were as follows: room-temperature and room-velocity were 285 K and 0.02 m/s, respectively, and initial air velocity inside the chimney was 0.28 m/s; we imposed a voltage of 80 V upon the heaters inside the closed space at 8:45 am and kept it constant. Supposing there was no energy lost from the periphery and bottom of the closed heating space, the heat flux from the bottom of the collector (two pieces of iron sheet) to the air was accordingly 213 W/m^2.

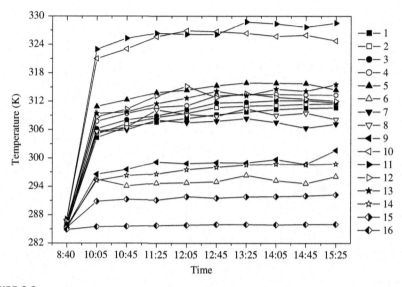

FIGURE 8.3

Variations of temperature with time on Jan. 13.

Measurement conditions for Fig. 8.4 were as follows: room-temperature and room-velocity were 281 K and 0.02 m/s, respectively, and initial air velocity inside the chimney was 0.15 m/s; we imposed a voltage of 120 V at 9:20 am and kept it constant, the heat flux from the bottom of the collector to the air was accordingly 480 W/m². The time interval for the measuring between Figs. 8.3 and 8.4 was 10 days, which was mainly for the purpose of avoiding the case that the former measurement had any influence on the latter measuring conditions and measurement results.

It can be obviously seen from Figs. 8.3 and 8.4 that the measurement results above indicate that the toggle speed of the system is quite fast. After heating for an hour, the maximum temperature increase for the chimney inlet air reaches 38 and 56 K, respectively, afterwards, the temperature increase for each measurement point is fairly gentle, and the system is basically under a steady state. Besides, through comparing Figs. 8.3 and 8.4, it also can be found that the effect of heat flux on the system temperature variation is very remarkable, the higher the heat flux, the more remarkable the temperature variation of each measurement point.

Measurement conditions for Fig. 8.6 were as follows: room temperature and velocity were 299 K and 0.08 m/s, respectively, and initial air velocity inside the chimney was 0.08 m/s; we imposed a voltage variation from 6:30 am till 6:30 pm to simulate the solar radiation which met the sine rule: the voltage imposed and the resulted heat flux from the bottom of the collector to the air are shown in Fig. 8.5.

Obviously shown in Fig. 8.6, the temperature variations of measurement points inside the system meet the sine variation rule. The maximum temperature

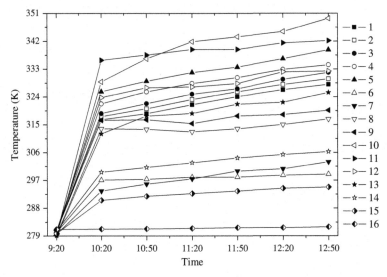

FIGURE 8.4

Variations of temperature with time on Jan. 22.

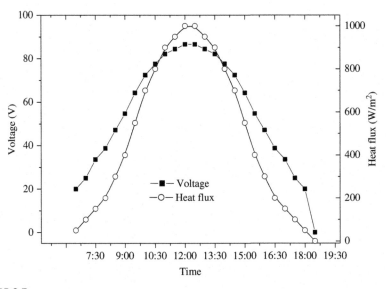

FIGURE 8.5

Voltage loaded on the system and the resulted heat flux on May 27.

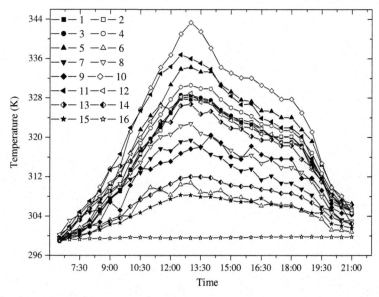

FIGURE 8.6

Variations of temperature with time on May 27.

increase reaches 43 K. And the largest heat flux occurs at 12:00, while the time corresponding to the maximum temperature increase occurs at 12:30. This indicates that the system has a lag effect with fairly short time, which shows that the existence of the closed space with heaters inside in the system has the characteristic of heat inertia. If a heat storage layer made of soil or gravel is used to take the place of the closed space in a commercial SC system, the characteristic of heat inertia of this kind of system will be quite notable.

8.4.2 VARIATIONS OF AIR TEMPERATURE AND VELOCITY IN THE CHIMNEY

Figs. 8.7 and 8.8 show the inner-chimney air temperature and velocity variations with time, respectively. Transparently, when the heat flux is 213 W/m^2, the maximum velocity within the chimney is only about 1.7 m/s; while the maximum velocity within the chimney is 2.9 m/s when the heat flux is 480 W/m^2. Experimental results indicate that under stable heat flux, the temperature increase and velocity within the chimney measured in winter are correspondingly larger than that measured in summer, and the experimental results agree well with the theoretical analysis results [16].

As shown in Fig. 8.9, when the heat flux at the bottom of the collector varies according to the sine rule, the temperature increase and velocity also varies

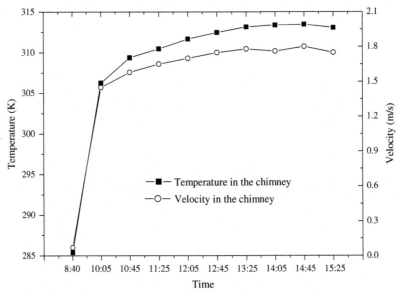

FIGURE 8.7

Variations of chimney outlet parameters with time on Jan. 13.

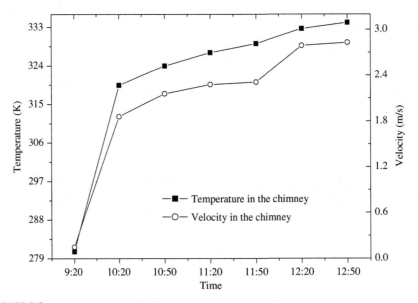

FIGURE 8.8

Variations of chimney outlet parameters with time on Jan. 22.

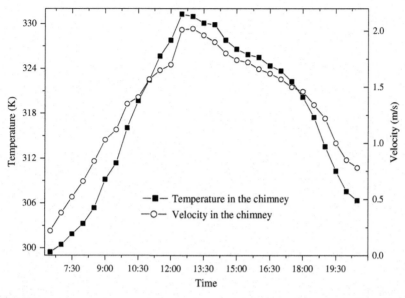

FIGURE 8.9

Variations of chimney outlet parameters with time on May 27.

according to the sine rule. And the air velocity within the chimney also reaches its maximum of 2.0 m/s at 12:30. Obviously, if a high heat flux is provided for a relatively long period (heating time is longer than 2 h with the heat flux larger than 500 W/m² before peak value is reached), and this velocity is far smaller than that in winter, the difference mainly lies in that when the temperature increase is identical and air is heated under a relatively high ambient temperature, the air density variation and the system buoyancy produced are relatively small, thus causing a relatively small air velocity. Therefore, under the same heating conditions, operation performance of the SC system in winter is better than that in summer.

8.4.3 TEMPERATURE DISTRIBUTIONS OF THE SYSTEM

It can be seen from Figs. 8.10 and 8.11 that the air temperature gradually decreases along the air flow direction within the chimney. Taking the thinness of chimney wall into account, the inner-outer temperature difference of the chimney and the heat loss through the wall are relatively large, hence the temperature variation within the whole chimney is quite remarkable and should not be considered as adiabatic, which differs from the theoretical analysis for a commercial SC power plant [14]. The main reason for this is that the theoretical analysis results are based upon the fact that the actual thickness of the chimney is relatively large, that is, thicker than 1 m, and the chimney is made of reinforced concrete, which

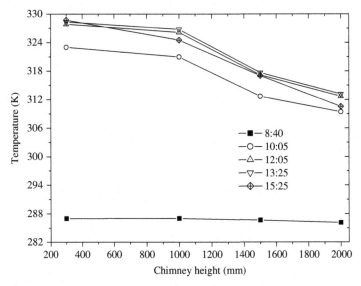

FIGURE 8.10

Variation of temperature in the chimney on Jan. 13.

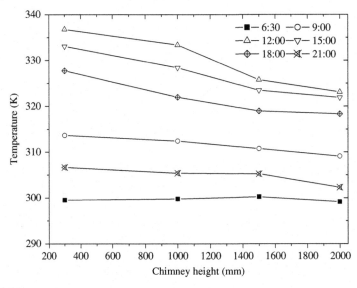

FIGURE 8.11

Variation of temperature in the chimney on May 27.

guarantees larger heat resistance and smaller heat loss. However, the chimney of the experimental setup in this paper differs greatly from the chimney of large-scale SC system, and as a result, analysis should be carried out according to the practical condition.

Furthermore, the experimental results also indicate that the temperature of the air flowing from the collector inlet to the bottom of the chimney is increasing all the time, the closer to the interior of the collector, the higher the measurement point's temperature.

8.5 CONCLUSIONS

Experimental research for a mini-scale SC prototype was carried out, and under the conditions of both constant heat flux and heat flux varying according to the sine rule, the air temperature variations within the collector, air temperature, and velocity variations within the chimney were all measured. Experimental results indicate:

1. The air temperature increase and velocity within the chimney in winter are both larger than that in summer, which agrees well with the theoretical analysis results.
2. Temperature from the inlet of the collector to its center gradually increases, which agrees with the theoretical analysis. However, the temperature varies remarkably along the air flow direction within the chimney, thus the airflow within the chimney should not be regarded as an adiabatic process, which is quite different from that of a large-scale SC system, and this is mainly caused by the thinness of the chimney wall, which means small heat resistance and considerable heat loss.

REFERENCES

[1] Schlaich J. The solar chimney. Stuttgart: Edition Axel Menges; 1995.
[2] Haaf H, Friedrich K, Mayer G, Schlaich J. Solar chimneys, Part a: Principle and construction of the pilot plant in Manzanares. Int J Sol Energy 1983;2:3—20.
[3] Haaf H. Solar chimneys, Part b: Preliminary test results from the Manzanares pilot plant. Int J Sol Energy 1984;2:141.
[4] Kulunk H. A prototype solar convection chimney operated under Izmit conditions. In: Veziroglu TN, editor. Proceedings of the 7th Miami international conference on alternative energy sources. Miami Beach, FL, USA: Hemisphere Publishing Corporation; 1985. p. 162.
[5] Pasumarthi N, Sherif SA. Performance of a demonstration solar chimney model for power generation. Proc. of the 35th heat transfer and fluid, Sacrmento, CA, USA. 1997. p. 203—40.

[6] Pasumarthi N, Sherif SA. Experimental and theoretical performance of a demonstration solar chimney model-experimental and theoretical results and economic analysis. Int J Energy Res 1998;22:443−61.

[7] Zhou XP, Yang JK, Xiao B, Hou GX. Simulation of a pilot solar chimney thermal power generating equipment. Renew Energy 2007;32(10):1637−44.

[8] Bernardes MA, Vob A, Weinrebe G. Thermal and technical analyzes of solar chimneys. Sol Energy 2003;75(3):511−24.

[9] Pastohr H, Kornadt O, Gurlebeck K. Numerical and analytical calculations of the temperature and flow field in the upwind power plant. Int J Energy Res 2004; 28(3):495−510.

[10] Denantes F, Bilgen E. Counter-rotating turbines for solar chimney power plants. Renew Energy 2006;31(12):1873−91.

[11] Ming TZ, Liu W, Xu GL. Analytical and numerical simulation of the solar chimney power plant systems. Int J Energy Res 2006;30(11):861−73.

[12] Ming TZ, Liu W, Pan Y, et al. Numerical analysis of flow and heat transfer characteristics in solar chimney power plants with energy storage layer. Energy Convers Manage 2008;49:2872−9.

[13] Ming TZ, Zheng Y, Zhou Z, Liu W, Pan Y. Unsteady numerical simulation of a solar chimney power plant system with energy storage layer. J Energy Inst 2009:1 In Press

[14] Ming TZ, Liu W, Xu GL, et al. Numerical simulation of the solar chimney power plant systems coupled with turbine. Renew Energy 2008;33(5):897−905.

[15] Ming TZ, Zheng Y, Liu C, Liu W, Pan Y. A simple analysis on the thermal performance of solar chimney power generation systems. J Energy Inst 2009:1 In Press

[16] Gannon AJ, von Backstrom TW. Solar chimney cycle analysis with system loss and solar collector performance. J Sol Energy Eng 2000;122:133−7.

Research prospects

Tingzhen Ming[1,2], Wei Liu[2], Tingrui Gong[2], Wei Yang[2], Dong Chen[2] and Zhengtong Li[2]

[1]*School of Civil Engineering and Architecture, Wuhan University of Technology, Wuhan, P.R. China* [2]*School of Energy and Power Engineering, Huazhong University of Science and Technology, Wuhan, P.R. China*

CHAPTER OUTLINE

Obviously, existing research on the technical aspects of the solar chimney power plant (SCPP) is quite numerous. However, there are still many issues that have not been entirely addressed. Key scientific issues associated with the system are shown in Fig. 9.1. The author of this book contributes only a small part to the whole work in this field research. Just for the purposes of analyzing the heat and mass transfer processes in the solar chimney power generation system, the author believes that at least the following points need to be investigated in further research.

9.1 THERMODYNAMIC THEORY FOR THE LARGE-SCALE SCPP

Compared with the conventional thermodynamic cycle system, there is a large-scale interface between the SCPP system and the environment. Transient atmospheric and terrestrial environment will have significant influence on the thermodynamic cycle processes. Heated air and the system energy conversion medium have complex flow characteristics, which will have significant effect on the efficiency of solar chimney system. There is no decisive breakthrough in the existing thermodynamic research; thus the system efficiency is low, the relative humidity of the working fluid will

Solar Chimney Power Plant Generating Technology. DOI: http://dx.doi.org/10.1016/B978-0-12-805370-6.00009-0

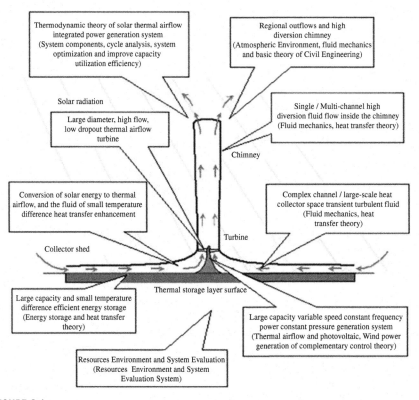

FIGURE 9.1

Basic theories for the large-scale SCPP.

further deteriorate the overall efficiency. As a result, the chimney should be built up to 1–3 km which is a big challenge for architecture and civil engineering. In order to improve the energy conversion and utilization efficiency, thermodynamic theory and energy conversation mechanisms in large-scale SCPPS should be carefully studied. In addition, innovative design on the SCPP will also be helpful to increase the system output. Decreasing the heat losses in the processes, researching the interactions between the atmospheric environment and the system based on the thermodynamic processes in different solar chimney systems, and exploring the calculation methods for thermodynamic cycles also need to be investigated.

9.2 EXTERNAL FLUID FLOW AND HEAT TRANSFER IN LARGE-SCALE CHANNELS

A remarkable feature of the large-scale solar chimney plant is the large-scale dimension, so the internal and external flow characteristics are significantly different from conventional sized channels. The existing theoretical models, technical data, and calculation methods for the large-scale may not be able to be adapted or may be even completely invalid. To this end, we have to conduct an in-depth study on the mechanism of energy conversion of the large-scale SCPPSs, based on relative theories on fluid flow and heat and mass transfer mechanics.

Meanwhile, we should explore the energy conversion process, analyze the coupling relationships between the various physical fields, and explore the turbulent models in complex flow channels. Further study should be carried out to research energy storage, release, and distribution in the energy storage layer. We should design energy storage systems and find power generation peak regulation solutions for different seasonal conditions. Finally, more work should be done to reveal the relationship between the structural parameters and the solar system—thermal energy—mechanical energy conversion efficiency.

9.3 TURBINE RUNNING THEORY FOR THE LARGE-SCALE SCPPS

The coupling mechanism of fluid and solid regions, airfoil type of the turbine blades, and flow channel design theory are challenging issues in SCPPS. The fundamental theory of unsteady flow, related research including dynamic balance in the large-scale thermal airflow turbine machine, unsteady flow mechanism inside the thermal airflow turbine, and unsteady trail interference between turbine blades need further research. Besides, more research should be conducted to explore damping methods for the blade torsional vibration and to study the blades abrasion mechanism caused by the wind and sand within the turbine. What's more, further research should be conducted to analyze the coupled heat transfer and flow characteristics in the storage layer, collector, turbine, and chimney, the four essential parts of the solar chimney system. Also, more attention should be drawn on the impacts of the circadian, the weather, and the changes in solar radiation on the coupling heat transfer and flowed characteristics within the entire system.

9.4 THE IMPACTS OF ENVIRONMENTAL FACTORS ON OF LARGE-SCALE SCPPS

Various natural environmental factors such as wind, storms, dust storms, earthquakes, and other work environment factors will exert a crucial influence on the reliability and life-cycle of facilities and equipment in large-scale solar chimney power system. Considering the chimney and collector's scale (the chimney height of MW-scale solar thermal airflow power plants is in hundred meters, collector diameter is in kilometers) and harsh environmental conditions in West China, comprehensive study should be conducted to analyze the physical basis, mechanical properties, and structural stability of the ultra-high solar chimney from many aspects, including the structure design, simulation, and virtual design, etc. We must study in-depth the excitation effect and coupled physical response of the external environment, research the structural shape characteristics and mechanical properties of ultra-high solar chimneys and large solar greenhouses, explore the interaction between environmental factors and system responses and anti-interference mechanism of system, provide a theoretical basis for the system to protect against ecological disaster, and improve system reliability.

9.5 NEW-TYPE LARGE-SCALE SCPPS

Though many researchers have proposed a variety of ecotype solar chimney power generation systems, the reports conducting an in-depth study of these new systems are rare. Most of these works only proposed a feasible plan. However, there has been no comprehensive study covering all aspects of the system including theory, technology, economy, and feasibility. More research should be conducted to explore the use of the solar chimney plant to form a new "Energy—Environment—Ecology" coordinated development model, and to establish the comprehensive benefits evaluation system. Utilizing the greenhouse of SCPPS to conduct fundamental research in agricultural facilities, we may grow vegetables, plants or agricultural crops in Western China. To achieve these, many problems should be addressed, such as developing agricultural facilities suitable for the local environment, getting investment, and achieving benefits. All the issues above are worthy of further attention.

If the energy problem is not solved, it will constrain China's economic development. In addition, it is an urgent task for researchers to find more fossil alternative energies to alleviate the severe ecological and environmental situation in China. Further attempts should be made to achieve environmental improvement and ecological reconstruction in China and worldwide.

Index

Note: Page numbers followed by "*f*" refer to figures.

Printed in the United States
By Bookmasters